An Ordinary World

The Role of Science in Your Search for Personal Meaning

Todd Duncan

Science Integration Institute
Beaverton, Oregon

An Ordinary World

The Role of Science in Your Search for Personal Meaning
2nd Edition

Published by:
The Science Integration Institute
3390 SW Lundgren Terrace
Beaverton, OR 97005
http://www.scienceintegration.org
info@scienceintegration.org

ISBN 978-0-9712624-1-6 (paperback)

Printed in the United States of America

Cover photograph is from PD Photo.org and is in the public domain at http://pdphoto.org

Contents

Acknowledgements

I would like to thank everyone involved in the founding of the Science Integration Institute and the development of the ideas underlying it, including Connie Barlow, Bill Becker, Terry Bristol, Kelly Bromley, James Butler, Devi Chandramohan, David Christian, Kim Coble, Bekah Dougher, Michael Dowd, Alley and Avis Duncan, Larry Edwards, Chris Fuchs, Brian Gardener, Anita and Paul Harris, Candice Holcombe, Daniel Holz, Jessica Jerome, Elaine and Tom Johnson, Claudine Kavanagh, Kelley Martin, Claire Masson, Doug McCarty, Bob McGown, Cole Miller, Amy and John Peters, Jean Quashnock, Andreas Rechtsteiner, Carly Riter, John Rueppel, Jack Semura, Daniel Sheehan, Murali Sundar, Jim Sweitzer, Brian Swimme, Susan Thomas, Craig Tyler, Aparna Venkatesan, Jack and Patricia Watson, Eric Weeks, Katie Whalen, Leslie Witterschein, and Martin Zwick. The ideas expressed here reflect the many conversations we've had in shaping and refining ways to make science more accessible and meaningful to a wider range of people. My deep appreciation to Amanda Duncan, for input on this project and for her partnership in all the adventures of life, and to Angela Lowman for her encouragement and assistance in the production and distribution of this book. Jacob Wilson assembled the original list of reflection questions that appears in modified form at the end of Chapter 2. In addition to their valuable conversations, Geoffrey Hamilton, Maya Lessov, and Scott Wunsch contributed helpful feedback on an early draft of the first edition. As always, I am grateful to my parents, Sara and Jerry Duncan, for their steady encouragement and support.

Preface to the Second Edition

One of my most powerful childhood memories is of watching Carl Sagan's PBS series *Cosmos*. Sagan's integrated portrayal of the multi-faceted human quest to understand the cosmos —weaving together its scientific, historical, philosophical, and artistic aspects—touched deeply my own emerging questions about the universe and my place within it. At the time I had little understanding of what it meant to be an astronomer or a physicist, but I was hooked: I wanted spend my life investigating the universe in the manner that Sagan outlined.

Two core themes emerged from my subsequent reading and pondering of the universe. I'll explain these themes because they form the foundation for this book and really all of my scientific work.

The first is a strong impression that there is something deeply wonderful and meaningful beneath the surface of our everyday awareness. The more I learned about the vastness and subtle interconnectedness of the cosmos, the more clearly it struck me that *we humans are an integral part of this grand tapestry*, and our lives have some deep meaning as part of its context. We may never fully grasp or articulate exactly how our lives matter within this vast context, but we can sense it during moments of inspiration. For example, looking up at a dark starry sky or listening to the musical background from *Cosmos* always send shivers of wonder up my spine, a sign of my unconscious awareness that I'm connected to something grand.

The second theme is a recognition that we have an extraordinary capacity for self-deception. It is a mysterious but obvious fact that we're capable of believing things that are not true, particularly when it's in our best interest for those things to be true. Wishful thinking is a powerful force, and we must guard carefully against being drawn into believing something just because we wish it were so. Science is important as an error-correction process, keeping us honest as we search for understanding of the cosmos and our place within it.

This book is about the apparent tension between these two forces: the desire for meaning and the desire for accuracy. For reasons I don't completely understand, it's easy to feel that a commitment to scientific accuracy requires an acceptance of the ultimate meaninglessness of human life. That impression is hard to shake partly because I'm aware of how much I *wish* my life to be meaningful. This strong desire puts me on guard, wanting to make sure I'm not carried away by wishful thinking that is contrary to the evidence. So I'm very vulnerable to interpretations of science telling us that the cold, hard facts require us to accept the meaninglessness of life.

This tension is related to the widespread belief that science is somehow dehumanizing, removing meaning and magic from the world and making it ordinary and dull. The title of this book comes from one of my early attempts to express in a poem this common impression that science turns a magical and meaningful world into an "ordinary" world:

An Ordinary Night

The sparkling of the winter moon
On the newly fallen snow
Is only a distant rock's reflection
On some frozen H-2-0.

And the stars so brightly shining
Against the moonlight glow
Are only spheres of plasma blazing.
That we certainly know.

Even the whisp'ring of the icy wind,
Which does so softly blow,
Is mostly moving nitrogen
As anyone could show.

The steam that drifts from the chimney tops
Is of course just condensing water.
But so softly it drifts before my eyes
In air so cold that it crackles and pops,
That it gives the scene a dreamy feel
Yet, somehow, makes it seem more real.

For the ingredients of this winter night
Are ordinary and plain.
But combined together exactly right
They become a magic scene.

In the following pages I investigate how to understand and reconcile the apparent tension between accuracy and meaning. I conclude that it is possible to maintain a balance between the two forces. We can guard against being seduced by wishful thinking, while also not mistakenly over-interpreting the science as telling us that no meaning is possible.

Since many people seem to share this conflicted relationship with science, I hope this attempt to articulate the problem and suggest a direction for solving it will help others as well. Please use the book as a starting point for your own explorations, and feel free to share with me any ideas and insights you discover along the way.

The approach here is intentionally very exploratory and open-ended. This book outlines a possible framework and a perspective for approaching science, rather than trying to teach any particular science or particular worldview. I'd like to leave the way as open as possible to a variety of interpretations. Other tasks are for other books. A much more detailed survey of our modern scientific understanding of the universe can be found in my text with Craig Tyler, *Your Cosmic Context: An Introduction to Modern Cosmology*. And for readers interested in my take on what a realistic worldview might look like, given what we know about the universe so far, I'm working on a book called *Looking for Meaning in the Modern Universe*, which I hope to finish within the next year.

Todd Duncan
info@scienceintegration.org
November, 2008

Introduction

Science has had an uneasy relationship with the human desire for personal meaning and significance to our lives. Yet science and meaning are also deeply intertwined. Any honest attempt to better understand the universe and ourselves cannot ignore the remarkable success of the scientific process—we want any meaning we construct to be on solid ground. Science has worked so well that it is difficult to deny it a central role in telling us about how the world really is. Even if we don't like some of what the science tells us, because it may conflict with ideas on which we happen to base our current sense of meaning, we feel obligated to pay attention and look to it as a source of information.

Evidence that many people sense this can be seen both in our fascination with science and in our reactions against it. The widespread interest in popular science books reflects a feeling that science must be relevant to the search for meaning. But scientific accounts of even the most fundamental theories of physics have a hollowness to them that always leaves us wanting. In one sense they profess to offer the whole story. But in another, more intuitive sense, we know they can never be the whole story because they leave out what is the very stuff of life to us. Scientific descriptions of the universe seem too far removed from our experience, with nothing that brushes against the universe as we experience it in everyday life, the one in which we make choices and laugh and cry and seek a place for ourselves and our thoughts. Science provides a mental map of the universe, but it is in many ways an unfamiliar and unhelpful map, without clear connections to the concepts we operate with in our immediate experience. Most glaringly, it seems to lack a clear "you are here" marker needed to place ourselves within the framework of the map and use it as a guide to the choices we face in life.

This situation may explain some of the vigorous resistance to scientific ideas and the popularity of views that run counter to scientific evidence, but address directly the daily concerns of life. They offer something we have a deep and real need for, which science seems not to provide.

> *We feel that even when all possible scientific questions have been answered, the problems of life remain completely untouched.*
>
> —Ludwig Wittgenstein[1]

We all operate within a framework of concepts that make sense of the world to us, which we use to formulate our goals, hopes, and dreams, and to seek ways to overcome problems and obstacles as we build our lives. Certainly the universe out there has much to say about all this, but it's hard to figure out what it says when the scientific description exists for us as a remote framework without clearly articulated connections to the concepts with which we operate in daily life. So we live in a disconnected state: abstract and evolving knowledge of the grand universe on one hand, and the immediate need for a guide to our individual choices on the other hand. How do we bring these together, so that we can guide our immediate choices from a perspective that is informed by and connected to the big picture?

These connections exist, but they are easily disguised, lost in the abstractions. In some way we remain detached from the scientific descriptions, still not quite feeling they tell us much about the essence of the world as we experience it. For sciences that describe distant places and distant times, the links can seem almost impossible to maintain. The connections must be consciously made, the insights from science explicitly appropriated into our day-to-day awareness of who we are and how we interact with the world. We can learn to think not of the scientific universe out there, far away and long ago, but right here, where we live and experience the world. The big bang, for example, happened *here*, in the little region of space we can now hold in our hands, as well as out there in regions that are now more than ten billion light years away. We are just now receiving the glow from a condition once experienced billions of light years away, but that condition was also experienced right here, a long time ago. Bizarre properties of electrons and atoms and photons described by quantum theory can seem abstract and detached, until we realize we're talking about us, too—our atoms, the air we breathe, the sunlight which sustains us. Many of the remarkable insights from science remain disconnected from our personal worldviews which are the maps we use to guide our choices and our lives. But this need not always be the case.

I suggest that the meaning behind our individual lives, which science can help us uncover, is not to be looked for only in regimes where our current scientific understanding is stretched or incomplete—in exotic theories of the early universe, in black holes and warped spacetime and the arcane mathematics of grand unified theories. It is found, rather, in the "ordinary world"—the world in which we live every day, but which really is so full of mystery and wonder that it seems inappropriate to call it ordinary. If we can learn to live with a full awareness of our connections to the universe we are a part of, the world may never seem ordinary again.

My goal in this short book is to illustrate that the seemingly opposing aims of "personal meaning" and "consistency with science" need not be in conflict. On the contrary, the process and insights of science can act as a valuable filter and guide to developing our sense of being part of a bigger context, within which our lives have meaning. We live our lives motivated and guided by a set of beliefs about how the world works and how we connect to it. These connections are all around us, in every action and every assumption about what is important for us to do. And science, while certainly not capable of providing all the answers, has a great deal to say about these assumptions. We just need to be aware of how to use the science, and what to look for.

My hope is that science can begin to play a much more central role in our varied individual efforts to construct an overall context for our lives. I hope we will learn to see new discoveries in basic science not as detached and esoteric curiosities, justified by the vague possibility of technological spin-offs, but as crucial steps in the process of uncovering humanity's role in the cosmos.

To help move us in that direction, I offer a point of view from which science can be seen as an important tool in your personal search for meaning in your daily life. Along the way, I also present some concrete suggestions for putting it to use for yourself.

Working Definitions

The following are not meant as formal definitions, but will provide an idea of my meaning in using these terms:

Science

> The word science refers both to a *process* for obtaining knowledge about the world, and to a *set of insights* about the world which have been built up through this process.

My working definition thus has two components:

1. Science is the process by which we invent possible explanations (theories) describing what we observe in nature, and then filter out explanations that work from those that do not work by testing (through experiment and observation) the predictions they make about what else we will observe. Two key features characterizing this approach are the important role of observation (disagreements are ultimately to be settled by experiment and observation—nature has the last word) and the search for unifying principles that attempt to connect many different phenomena with as few explanations as possible.

2. Science also refers to the body of knowledge produced by this inquiry process.

Personal worldview

The mental map by which we each view our relationship to the world, and which guides our choices and actions through the perspective it gives us on our individual role as part of the universe.

Science integration

The process by which scientific insights are incorporate into a person's personal worldview.

Meaning

A context within which our choices and actions are significant, so that what we do truly matters in some way.

This book is intended as an introduction and a gateway to your own exploration of your universe. Ongoing discussion and suggestions for putting the ideas expressed here into action in your everyday life can be found on the website of the Science Integration Institute at http://www.scienceintegration.org.

Chapter 1

Worldviews: Placing Your Actions in a Bigger Context

Summary: Our sense of what matters in our lives is grounded in a context of beliefs and assumptions about what the universe is like and how we as individuals fit into it. This context is a driving force behind most of our actions, so becoming more aware of it, and of the assumptions we make in constructing it, will have a direct and powerful effect on our everyday lives.

Our "mental models" determine not only how we make sense of the world, but how we take action.

—Peter Senge[1]

Most of the time, our awareness includes only a tiny fraction of all that's happening in the universe. We are naturally focused on the pressing concerns of our personal interactions with immediate surroundings. It's hard enough just to stay aware of the concerns of other *people* in our lives. It is even more difficult to back away and grant any tangible, direct, and immediate reality to the *overall* framework within which our individual lives are situated. Still, we are certainly aware that we're not individually responsible for our own existence. Our existence now is a result of events and processes that extend through space and time far beyond our immediate awareness; almost incomprehensibly far beyond, as modern cosmology research has enabled us to learn. Your body is like a fossil record of the entire history of the universe. The pattern of matter and energy that forms your identity as you sit there reading these words is an expression of a much vaster pattern spanning billions of years and including the entire ecosystem of Earth that nourishes and sustains us, the nuclear reactions in the sun from which light pours out as the driving force behind all life on Earth, ancient stars whose deaths produced some of the elements that now make up our bodies, and earlier cosmic processes that created the environment in which these stars could form. What you're doing and what you're thinking right now is not just you; it is a vast web of which you form the consciously aware part.

Though we may be largely unaware of our connections to a broader perspective, a little probing of our ordinary choices and actions reveals that they are expressions of an underlying "mental map" which incorporates our beliefs and assumptions regarding this universal context. Expressed in the form of this map, these assumptions have a direct and powerful impact on how we live our lives. In fact, the shape and direction of human society is largely a collective reflection of the mental maps or personal worldviews that encapsulate what the world is like, and how we relate to it, for each of us. The pervasiveness of these personal worldviews in guiding our lives and society increases as we become more technologically advanced, more able to shape our surroundings according to our mental maps.

We have the capability now for a society in which nearly everyone could have what he or she needed.[2] Many people are currently not doing work to directly produce things that are necessary for our survival. Entertainment, news media, fashion, advertising, banking, writing, art, much of science; none of these seem essential. Thus it is clear that we as a society have a great deal of time for non-essential things. If used properly, it seems we have the power to create a nearly ideal society. This was the dream of the industrial revolution and the age of technology. It would free us from the more mundane tasks involved in staying alive, freeing us to do greater things.

But what we do with our time beyond survival is all about our worldviews.

However vaguely, we all hold ideas about where we came from, where we are going, and how our decisions interact with the external world and its rules of operation to help or hinder progress in some direction. To see concretely the power of these worldviews, we need only notice that the dominant, driving feature of our lives is the need to make choices. Life presents to us an incessant, urgent demand to choose from an overwhelming array of options. We could never act on even a tiny fraction of all the options available to us as choices. Consider right now all the things you realistically could do: You could move your left hand, or your right hand, or shake your head, or throw this book in the trash (please don't!), or quit your job, or make a donation to the Red Cross, or read a book on surgical techniques, or sign up for a class on computer repair, or get on the internet and buy just about anything you could imagine (and afford), or...well, you get the idea. All of these choices and countless more are immediately available to you, right now, and many could significantly change the course of your future life and even the future of society. It's both empowering and overwhelming to become aware of this.

From all of these possibilities, what guides you into the one you actually choose (reading on to see what I'll have to say next, I hope)? Think about some of your more significant choices. Why did you make those choices, rather than the alternatives? What framework of beliefs and assumptions were you following? What gives you a sense that it makes a real difference which choices you make, so that you sometimes agonize about what to do?

For most things in the universe, it apparently makes little sense even to ask these questions. Most structures in nature simply do what they do, automatically and blindly following the orders that are somehow built into the fabric of space and time in which they are embedded. When an apple falls to the ground, it has no choice in the matter. We don't think to suggest that a wiser apple might have stayed on the tree a little longer until it was a bit redder and juicier. The apple doesn't feel guilty for hitting you on the head as it falls. Conscious, self-aware creatures like ourselves, on the other hand, have the unique predicament of feeling faced with choices about what to do. So our actions, rather than being guided only by the direct and automatic instructions of nature, are also guided by a mental map or worldview that *represents* the world in our minds. Strangely, we have the ability to create a wide variety of maps, independently of how things "actually" are, many of which conflict with the maps other people hold, and may even conflict with our own previous maps. And the map we hold right now may cause us to do things that are helpful or destructive to the order of things, depending on how well the story we are telling ourselves is in harmony with what's really going on in the universe around us.

> *In human affairs an idea is a greater moving force than any physical influence...So the shape of our future will depend to a large extent on our understanding of our role in the cosmic process.*
>
> —Louise B. Young[3]

We deliberate and agonize over important decisions because we believe the choices we make have significance on some basis. The foundation for the significance of our choices is some kind of overall context we are contributing to, within which it makes a real difference what we do. Of course, our immediate idea of this context may not be coherent or consistent or even fully conscious, and it may even contradict some of the abstract facts that we know about this context. Our assumptions about it can change from moment to moment, and the immediate personal worldview within which we act may often bear little resemblance to the one we would articulate in a more thoughtful, reflective state of mind. My point for now is simply that we *do* hold in mind such a context, even if only a makeshift one that serves for the moment. However we formulate it, this personal worldview is always operating in the background, and has a profound effect on the decisions we make.

To see this effect in detail, let's take an apparently trivial example. What factors influence my decision process when I go to the store to buy a new pair of running shoes? Well, obviously I want something that is comfortable and fits my feet. I try on a few pairs of shoes, walk around in them, jog a few steps. So far I'm just dealing with a direct and tangible response to what feels good. Then there's the latest information I may have picked up from *Runner's World* about the newest shock-absorbing materials, the best design for the type of running I do, and past injuries I'd like to avoid repeating. Here I'm anticipating future consequences for how I'll feel as a result of my choice. But there's much more thinking going on beneath the surface. Maybe I look for a shoe that's made in the United States, or one that I know uses environmentally-friendly manufacturing techniques. This last preference opens up a whole new set of assumptions. How did that automatic connection between environmental awareness and righteousness form in the first place? How do I know what will hurt the environment? Why do I think resources are limited and need to be conserved? Conserved for what? What sort of long-term value is behind this simple reaction? These thoughts are only the tip of the iceberg of what's really going on in a situation as simple as buying shoes.

So with nearly every choice, you make assumptions about the nature of the universe you live in. If you believe you should recycle materials, run a government in a certain way, reward certain actions and punish others, encourage or discourage the growth of technology, follow certain morals and goals in life, or even just buy a particular brand of shoes, the foundations of these beliefs hinge on certain assumptions about what the universe contains and how it basically works.

The automatic way in which we draw on our personal worldviews can also be seen in the familiar experience of chiding ourselves for wasting time. To be able to say with any conviction that you are "wasting time," you implicitly assume there is something more valuable you could have done, some purpose toward which your time would be better spent. These beliefs say something about your perception of the universe you are immersed in, even if you never articulate this perspective.

To further develop this idea in connection to your own life, pause to reflect on your activities of the past week. Think about your actions, the decisions you've made, the conflicts and struggles you've found yourself engaged in. This illustration works best with topics you feel most passionate about: the environment, equal rights and fairness, someone you love, a political philosophy or point of view, your career, your community. What really gets you out of bed in the morning and makes you want to face another day? What drives you to volunteer for a political campaign, or incites you to write a letter to the editor of your local newspaper? What ideas do you feel so strongly about that you'll argue them vigorously with others, despite the emotional strain this kind of conflict often entails?

As you continue to probe these questions, once again you'll come up against a complex set of assumptions and beliefs about what is important and what makes the universe tick. For example, suppose you think watching television is wasting our time and harming our society. This belief may cause you to devote time to educating people about the dangers of television, or maybe to avoid owning a television yourself, or simply to make an occasional comment about all the garbage on TV. But to presume that watching television is a negative influence, and to believe it is worth fighting against in any way, is also to hold assumptions about what is important fundamentally, that we would more appropriately spend our time on. Without such a background of beliefs, your argument ultimately runs out of steam, as purely a matter of personal preference, of no greater import than if, for example, you happen to like apple pie while I prefer cherry pie.

Similarly, if you believe that natural products are better than artificial ones (or vice versa), you must have some set of beliefs about what is basically going on in the universe that makes one more valuable than the other. What properties of natural products are different from the properties of artificial substances? What beliefs about the order and harmony of nature, and the extent to which it is appropriate for humankind to fiddle with this order, are behind your preference?

Try continuing this thought process with several important beliefs you hold. Ask of everything that you feel inclined to do: "What does it say about what I think is important, that I want to do that? Why do I think certain actions will produce the result that I want? What core assumptions are behind my choice of friends, career, where I live, what political party I support, what products I buy, what charities I contribute to?" Ask what beliefs or assumptions about the universe are necessary in order for you to have a solid grounding for any particularly strong opinions you hold. In some cases these core assumptions are well thought out and consciously made. In other cases, it can be very enlightening to make these assumptions visible to inspection and consideration for the first time.

The most fundamental conflicts we encounter with other people can often be understood in a clearer light by considering them as conflicting personal worldviews. When you fundamentally disagree with someone, it means that as they see the world at the moment, your purpose runs counter to theirs. Political and religious disagreements, for example, are so heated because they are deeply rooted in personal worldviews. The stakes are high, because the other person's point of view works against your most deeply valued objectives. It's not simply personal taste you're arguing about. You're disagreeing over the fundamental nature of the universe and what ultimately matters, how your life gets meaning in a broader context. Your own internal conflicts on such matters similarly reflect the high stakes involved.

The point here is not to decide whether you are right or wrong for believing in the ideas and causes you support. I'm only suggesting that you can benefit from *recognizing* the vast web of beliefs and the assumptions that lie behind them. This recognition opens the way for thinking about how you can best gain information about the world and your place in it, and how this information is incorporated into the perspective from which you make your everyday choices.

Your beliefs, whatever they are and wherever they originated, are based on some kind of information about the world. You have been collecting ideas and insights all your life (and much longer than that, if you consider the preferences and impulses built into your genetic code), to construct your current personal worldview; your version of how to put that information together in a meaningful way. As you gain new information, new insights, your perspective can change and evolve. In the next chapter, I'll be suggesting that we could benefit from using some of the tools and insights of science as part of this process that we're always engaged in. We have learned some incredible things about the context within which we live our lives, and our worldviews suffer from a limitation of perspective if they are formulated without an awareness of some of these insights. For now, I just want to point out and establish clearly that we are all involved in this process, constructing an operational "meaning of life" for ourselves, whatever tools we use to do so. We all live as if there is some background purpose behind our actions, but we rarely articulate our sense of this purpose and try to critique and clarify it.

Operationally, the web of beliefs that make up your personal worldview is the meaning of life for you, at this moment. I mention this because the systematic pursuit of the meaning of life is often joked about as a waste of time, an unanswerable question. It's a detached, philosophical question that one thinks about every once in awhile, if at all. But in a very real sense we *always* have an answer to this supposedly "unanswerable" question. We could not hold an opinion or make the decisions we make every day, without reference to at least a makeshift, temporary, and perhaps unconscious idea of what matters, what fundamental purpose we are working toward. Our motivation, sense of direction, and will to keep moving through life come in one way or another from the personal worldview that is our mental map of our world at the moment. And since we live within the context of our current personal worldview, anything that changes it will have a profound impact on our lives and on our society.

Given this, it seems that we would each benefit from becoming more conscious of the elements of our personal worldview, and from making a systematic effort to expand and to refine it. We know that our picture of the world is incomplete and that we can hold mistaken beliefs. We learn, discover mistakes in previous ways of thinking, change our minds all the time. Improving the accuracy and scope of our personal worldviews could thus directly affect the clarity and fulfillment of our individual sense of purpose. There are many possible ways to do this, but we first must be fully aware that we have personal worldviews, and then give some thought to the inputs that shape them. Later we'll look in more detail at the role science might have to play in this process.

The first exercise at the end of this chapter will help prepare the way for this, by continuing the process of making you more aware of your current personal worldview. It is intended to help you orient yourself within the universe *as you currently perceive it*. This will give you a starting point as you continue through the book, and will make your thoughts more conscious, for you to consider in the next chapter.

Reflection and Discussion

• **Describing Your Universe** – This activity is intended to make you more conscious of your own current beliefs about how the universe is set up, what properties of it are important, and how you fit into your own vision of its framework. This will help you to focus your thinking as you read the rest of the book, to be on the lookout for connections that are most meaningful to you.

> *Describe, as carefully and clearly as you can, what you think are the essential properties and features of the universe you live in,*

and what you see as your role in this universe.

It's best if you simply describe whatever comes to mind as important, but here are some possibilities to consider if you are stuck on what to start writing about: Imagine what you would see if you closed your eyes and floated out away from Earth. What would you see as you moved farther and farther away? How big is your universe? Does it have an edge somewhere? How is it arranged? (For example, are things spread uniformly throughout, or are some parts of your universe very different from others?) What guides the processes that happen within it, and makes them occur as they do? Has it existed forever, or if not, how old is it? Are there other planets around other stars? Are there other creatures besides those on Earth? What are the most important laws that control what happens in your universe? How do humans fit into the scheme of things? Are we important or not important? Do we have a specific role to play?

The idea is simply to spell out your own personal worldview in a concrete form that you can refer to. Have fun and see what you can discover!

• Consider whether the following passage seems true of your own life:

> *Man's main concern is not to gain pleasure or to avoid pain but rather to see a meaning in his life.*
>
> —Viktor Frankl[4]

Chapter 2

What Does Science Have to Do with Your Worldview?

Summary: From the many stories we can develop as worldviews to guide our actions, we need ways to filter what "works" from what does not, in order to clarify our goals and achieve what matters to us. Science can be viewed as the process and accumulated set of principles we have developed for dealing with the constraints of the external world and choosing what works over what doesn't. So it naturally has a role to play as a filter and guide in our search for a meaningful context to our lives.

Science is...one of the most important bases for meaning-making in today's world. The meaning drawn out of science by each individual who treads this path is a constructed, but not arbitrary, product of the human imagination. Despite the inherent subjectivity, meaning-making is not mere fabrication. It is a response to, a declaration of relationship with, Earth and the cosmos.

—Connie Barlow[1]

I hope you're now more aware of the key elements of your personal worldview, and see that this worldview plays a central role in how you live your life. But you may also still be wondering, "What possible relevance does science have to this?" One's personal worldview is, by definition, *personal* and subjective and is ultimately about *meaning*. Science is universal, objective, and has nothing positive at all to say about meaning. Right?

The need for something *like* science in our thinking is an expression of the constraints we all know exist in the world—a state of affairs that long predates the formal field of study we now call science. As we go through our individual struggles to give meaning to our lives, to give context and significance to what we do, or even just to survive, we always come up against the necessity of relating to the environment around us. To get anywhere at all, we must make our goals and our way of thinking match in some way with an external world that imposes itself on us whether we like it or not. The process of constructing meaning for ourselves does not occur in isolation; it is carried out in interaction with a world external to our minds and thoughts. We know from hard daily experience that we can be wrong about many things. Goals that we hold in our minds many times do not come to fruition in reality. Often what we believe is true, or believe will work in achieving our goals, doesn't work in practice. We need a method for filtering what works from what does not, in achieving what matters to us.

With the emergence of conscious self-awareness and the sense of choice discussed in Chapter 1, humans gained much greater power to plan ahead, to organize parts of the universe in new ways. But paradoxically, with this self-awareness also comes the power to apparently do other than what is "natural." Now our actions are not guided solely by automatic instructions from nature, but also, more and more, by our own mental map of the universe and how we fit into it. We make mental maps to guide our choices, but these maps may or may not represent directly the "real" universe as we might see it if we could step back and take the broadest possible perspective on what is going on. Rather, they reflect the immediate surroundings in which we formulate them: our society, the culture of our profession, our local economy, the isolated and artificial surroundings of our city or community. The experiences from which we build our mental maps may be very cut off from the overall process that is making our existence possible. So it's reasonable to ask how well we are living in the real universe.

This is the key point of contact, I think, between the perspective of science and the immediacy of daily life. I am a product of things I did not choose, did not have control over. Yet I am immediately aware of having choices and of the need to make decisions about what to do next. I feel a strong need to make these choices be in harmony with whatever context is behind the forces that brought me to this point, the point where I exist and am conscious of having choices to make. So I feel compelled to try and figure out something about the overall context, in order to make decisions that are true to it.

If I had *no* control over *anything*, then I would not care so much about knowing the overall context of nature, because I wouldn't need it in order to make good choices—they'd already be "programmed" in. On the other hand, if I had *complete* freedom, I also wouldn't care so much about understanding the context. In that case, I could make my own context, without feeling tied to an external one. I wouldn't need to know anything about the rules or context of nature if I were totally free to make up my own rules as I went along. But I know that I do not always have the power to do that. Choices I make now limit my future options, in a way that I seem to have no power to escape from. I must work within constraints that are externally imposed on me.

So we're caught in between these two extremes: uncertain and free to choose, but also aware that we're a part of something very important that we did not set up, to which we feel obligated to remain true in some way.

The rest of this chapter will develop the idea that the methods and insights of science can play the important filtering and guiding role in the process of developing your personal worldview and establishing a network of beliefs that give meaning to your life. To see that this is possible may require a different perspective on science than one with which you are probably familiar. I don't claim that this is the "right" way to look at science, or that it is necessarily the way most scientists view it as they go about their work. But I do think it is a perspective that has the potential to change the relationship between science and society, and to make science a more meaningful and integrated part of your life. I offer it simply as one perspective that might prove meaningful and useful to you. This perspective has two key components:

1. The subject we call science has emerged through the process of trying to work with or overcome the obstacles imposed by the external world around us. Nature has built-in constraints and limitations on what we are able to do. These constraints hinder our efforts to achieve our goals; or more accurately, they define the ways in which we have to work to achieve our goals. We've developed science as a direct response to our recognition of these constraints that are external to our immediate thought processes. We can't wish that we were on the moon and suddenly find ourselves there merely by thinking about that goal. For some reason, nature does not work that way. However, there *is* a way we can get to the moon, provided that we follow a general sequence of steps (which involves building a rocket, obtaining suitable fuel, understanding and applying the laws of physics that determine the trajectory of the rocket, etc.). Through science, we try to understand exactly what the constraints are, and figure out how to achieve our goals within those constraints. The precision, skepticism, and attention to detail that

characterize science exist because getting the "right" answer (*i.e.* the one that works) really *matters*, if we want to achieve our goals. So, to the extent that our ideas about meaning and purpose cannot be separated from the external reality of the world in which we live, and from the constraints this world places on our actions, these ideas also cannot be separated from science.

2. Insights provided by science can be assimilated into an overall picture of the arrangement and history of the universe, including (among many other things) the emergence of life and consciousness. To a large extent, our efforts to give meaning to our lives amount to developing ways to place what we do within a bigger context. This process is relatively familiar when described in terms of human culture. We're well aware that much of the meaning and significance we attach to events in our lives is embedded in the history of our society through our cultural beliefs and practices. But less familiar is the awareness that this idea can be extended to a much broader and more fundamental context. We arose as part of a process that has occurred over a tremendously long period of time, much of it long before life of any kind was present on Earth. Certainly most of this process occurred without our control, consent, or choice. So to understand anything very fundamental about what our lives mean, we need to know something about this process: its key stages, the rules that seem to guide how it works, and so on. And of course even as we work to understand this process, we are in the midst of it; the process certainly did not stop with the emergence of humans on this one planet.

The combination of these two components suggests that we make use of science as a process by which we gain the capability to do things that fundamentally matter to us. Seen from this perspective, the enterprise of science gains a new relevance and a new connection to our individual struggles to do something meaningful with our lives, and it gains a more clearly significant role in constructing and living our personal worldviews. In one way or another, all of our major concerns are connected to the underlying question, "How do I gain the capability to do the things that matter?" Everything we do with any passion, from the basic search for food and shelter, to the most esoteric branches of philosophy or art, owes its basic significance to a connection with this underlying question.

Think about some of the times you've felt particularly excited and enthusiastic about your life, and contrast these with times when you've felt uninspired or unmotivated to do anything. What seems to motivate us and make us eager to push forward with life is a concrete awareness of something fundamentally *worth* doing, which is also within our *capability* to do. If we agree that life is mostly about gaining and exercising the capability to do things that fundamentally matter, then we can see why the insights and methods of science are directly relevant to the search for meaning in our lives. Finding meaning involves identifying what really matters to us, and then figuring out how to arrange things so that these "things that matter" actually come to pass. By providing insights into how the world works, science is essential to both of these components: It gives us the perspective to see more clearly what things really matter, and helps us gain the ability to actually do those things.

A general characteristic of the scientific enterprise as it is often portrayed and perceived is that the process of "gaining the ability to do things" is kept mostly separate from the question or the search for "what we are fundamentally trying to do." I'd like to encourage a closer connection between these two aspects of our experience, in a way that will make science more directly relevant. To begin this process, the next few pages will be spent demonstrating in a general way how science can be applied to elements of your personal worldview, and will serve to illustrate the questioning and thinking process as it might proceed from this perspective. The next chapter will apply this process to some specific examples from Chapter 1.

So, let's start with our own thoughts as we sit here, reading, thinking, or writing. As you step back and look at your thoughts, notice that you feel a certain way about your current situation, and also probably have certain wishes about how you would like things to change for the future. As you begin with a reflection on where your satisfaction and sense of meaning and direction comes from, you see that it is mostly a matter of your state of mind. This suggests the follow-up question: Why do we need the external world at all?

Of course you can immediately see that external conditions do have a great influence on your feelings and state of mind. You may know that you feel a certain (good) way when you have just eaten a nice meal, listened to your favorite music, traveled to your favorite location to watch a sunset, or achieved some important personal goal. But for the most part, you cannot produce the same state of mind merely by wishing it so. Life puts constraints on us, requires certain external arrangements in order to produce the internal feelings we seek. Not everything we believe matches reality; we can't do things just by wishing them true. As you continue to think deeply about your personal worldview, a key concept that becomes apparent is that you have control and choice about some things, but many things are imposed on you externally, and are beyond your control. An important feature of our experience is this interplay between our wishes and ideas about how we would like things to be, and the constraints and limitations that are imposed on us by an external world.

One example of this limitation is the impact of our surroundings on what we think about. To take advantage of this, if possible, I suggest you read the rest of this chapter (with a flashlight!) out under a dark, starry sky. If that's not possible, do whatever you can to set up an environment that encourages you to think about the big picture and your connections to the universe we live in. (The reflection questions at the end of this chapter may help with this.) The connections between science and the questions you ask yourself in deciding how to live every day emerge much more easily when both aspects of the universe are right in front of you at the same time.

Now we're ready to think more about the external world and how we interact with it. The starry sky is an awe-inspiring sight on a clear, dark night. In addition to perhaps making you feel small and insignificant, it fills you with a sense of wonder. It confronts you, point blank, with the realization that so much lies out there beyond our immediate reach, even beyond easy reach of our thoughts. Maybe there are places we cannot even begin to comprehend, where things are completely different from the world we see here on Earth. As you look out, might there be some other creature, on some other planet in your line of sight, looking back toward you and wondering, like you, if anyone else is out there? The immensity of the night sky compels you to think beyond your day-to-day concerns and ask questions about what the universe is doing here and how we, on our planet, fit in.

Let's first just ask some of the questions prompted by what we see. What would we most like to know, if we could know anything? We each have our own ways of phrasing them, and have different emphasis, but I think the general idea is usually the same for most people when they stop to think about it. So, I'll give you my own versions of some of these questions to get you started and to get you in the right mood to think about your own. A good analogy to keep in mind as we go through this process is to imagine finding ourselves on a stage for a very small fraction of the total performance time of a play. So we have a brief chance to look around at the way the stage is set, at the costumes and props, and maybe hear a few words of what the actors are saying (perhaps in a language we don't understand and must translate). We can use our observations to figure out something about the plot of what is happening in the performance, and to figure out what role we could play in it, if any. After all, if we're there on the stage, during the performance, maybe there's something we're supposed to do, some part we play, that we don't know about yet. Here, of course, the "stage" is the entire universe.

Looking around, I realize that there is an incredible amount I don't know. How far out into space could I go, and still find stars? Are things the same, way out there, or are they completely different from what I see here on Earth? (For example, do stars work the same way our sun does? Are there other planets made out of the same elements, and the same kinds of rocks, as our Earth?) But even more disconcerting than not knowing about so many things way out there in space is how little I know even about myself. Where did I come from? I can only remember a few decades of history, yet people tell me that they can remember things that happened before I was born. History books tell me that there were entire civilizations long before ours, which thrived for thousands of years in some cases. If I doubt the words of these books, I can find more direct evidence that things were happening hundreds, thousands, or more years ago. There are artifacts, remains of the great works of past cultures, which lend support to the idea that these societies really existed. I can use this evidence to put together a history and a mental picture of what went on in the past and how some events led to others.

The same sort of evidence, though sometimes more subtle, exists for events in the history of nature. The rings on a fallen tree indicate that it survived hundreds of cycles of the seasons before falling to a storm or a saw blade. The craters on the moon look very much as if they are the result of objects that have smashed into its surface at high speeds. But if this is true, many of these impacts must have happened millions or billions of years ago. A large asteroid that probably hit the Earth 65 million years ago left behind trace elements that give away its presence and may reveal some of its secrets. The history of how the solar system must have formed 4.6 billion years ago is expressed in what the solar system looks like today. The composition of stars reveals something of their history. If we know how to interpret the evidence, we can put together a history of the entire universe, just as we can put together a history of our society. We often forget that this universal history is *our* history; it describes the events that have made us what we are, just as surely as the more recent and accessible history of the society we live in tells a story of how we came to be as we are today.

So where was I when all of these great events were going on? Perhaps a more appropriate question: Where was whatever was to become me? What was it doing to prepare itself, to make it possible someday to become me? Did it know what it was doing, that it was progressing in organization and complexity? Of even more pressing importance: What will happen to the stuff that makes up me when it is finished being me?! After I have returned to "dust," will the dust keep any memory of the fact that it was once not "mere" dust, but rather the remarkable structure that was me?

Whatever the answer to the last question, there is still much I can do before that happens, while I still definitely am me. But what should I do? I am fortunate to live in a place and time at which I have great choice about what to do with my life. What I really mean by this is that I live in a circumstance in which many of the needs that are externally imposed upon me, as necessary for my continued consciousness, can be met with relatively little extra effort on my part. As long as I can find some way to make a living (by carrying out actions that allow me to tap into a system that can provide my needs), I can do as I please. Should I be spending my time writing this book? Should I spend most of my life in quiet contemplation and thought, trying to understand how I came to be here and trying to ignore physical concerns as much as possible? Should I follow a life of action, in which I work to change things, to protect the environment of our planet, or to build better conditions for people whose basic needs are met only with great struggle? Or what about those stars out there? Should I peer through telescopes to try to understand more of what is out there, in the hope of finding where our human destinies lie? Or should I work on building rockets, designing spacecraft and propulsion systems that can take people out to discover what is there, instead of only looking passively from far away? Or should I stop trying to explore further and find out more, and instead believe that the answers to how we should live are already known to us? Should I strictly follow the teachings of one of the world's great religions, believing that if I am successful in following the rules, I will have helped accomplish what mankind is supposed to do? But then, there are many religions, with different beliefs and different rules. Which one should I choose?

All these questions seem to be driven by the feeling that we want what we do to matter, to have some context in which our actions are grounded. Let's try to trace through a little of how our thinking develops on this.

Early on we learn that some things are "worthwhile" to do, and some things are not, that are considered to be a "waste of time." Our feelings that we are doing well as opposed to goofing off or wasting time are based on this. In school we should focus on our studies and work hard, rather than causing trouble, daydreaming, or playing games. Studying is usually considered superior to watching television, for example. But what is it that distinguishes these categories? Maybe it's a matter of what is most beneficial in the long run. Playing games may be more fun now. But we gradually learn that if we forego immediate rewards, we may be able to get much greater rewards in the future. I may not feel like memorizing names and dates for a history test tomorrow. But if I do this now, I probably will do well on the test, and will feel good about it then.

Now that I have begun to realize it is better to look at things that accomplish something worthwhile, in the long run, I can continue to look on a higher level, further ahead. I begin to wonder, "Why does it really matter that I do well on a test?" I feel good about it; I may receive praise for it. But why do I feel good? Why do I receive praise? What is it that makes this more worthwhile than, say, having spent the time playing soccer instead of studying, and then getting a poor grade on the test? (I love history, but must admit that playing soccer usually sounds more tempting than studying for a history test!)

Perhaps we assign more importance to longer-term goals. It is not really just my performance on the test that is being rewarded (though the deeper basis for the reward may be long forgotten in many cases). If, in the process of preparing for the test, I really did learn something about history, I will be better able to contribute to the development of my society in the future. I will have a better understanding of the past, of how we came to be where we are today, and so may have a better view of where we should be going in the future.

This suggests that we should move the feeling of wanting significance for things up to the level of the society. Here it may be easier to see why a view for the long term is important. We have made great progress in gaining for ourselves the material freedom we have today. We can, for example, hop in a car and drive nearly anywhere we wish to go. We enjoy this freedom, and as individuals want to take full advantage of it. However, without any restraint on its use, the society would quickly lose the benefits of it. We could quickly run out of the fuel that makes automobile travel possible, or we could damage the environment sufficiently to make other aspects of our lives less pleasant. We must put off some of what we would like to do now in order for progress to continue in the future. Once it occurs to you to really see and think about the distant future, a whole new world is opened. You begin to think not just of what you want right now, but why you want those things, and what long-term significance trying to get them will have.

But once again we can ask why our concern for the future matters. Why do we believe in progress at all? We consider it important that we conserve resources so that our planet's ecosystem is not badly damaged within the next century. But it's clear that our planet will not last forever in its present form anyway. Change is an inherent part of nature. We have only a limited supply of resources on Earth, so no matter how conscientious we are, if we are converting some resources into others, we cannot continue that same process indefinitely. Our sun is gradually changing its composition, and someday will not be able to sustain life on Earth through the processes it carries out now. What are we working toward? Is there anything that will have a permanent significance? We somehow feel that there must be (this feeling seems to be the origin of our interest in long-term benefits of our actions), but it's very difficult to understand how or what form such permanence would take.

If there is some purpose that we ought to try to carry out, what will be the result if we find and carry out that purpose? Can it have any permanent, long-term significance? After we accomplish it, then what? Does the universe reach the state of harmony it has been seeking, and remain that way forever? But what happens during this "forever?"…

Before we drive ourselves completely crazy, let's pause to notice that the act of fighting against odds and struggling to accomplish things seems inseparably linked to any notion of purpose. Somehow it does matter that we have goals, and that we have obstacles in the way of our goals, so we can work to achieve them. It must be subtle, whatever the resolution, and we are probably still a long way from finding it. You probably have some ideas about this but, like me, are not sure where they lead or where they can end.

Are we to conclude from all this that the task is hopeless, that it doesn't make sense for there to be significance to anything that we do? I don't think so. The very fact that we feel a desire for meaning must come from somewhere. Whatever it is about the universe that makes it possible for us to want meaning seems likely to also provide possible ways to meet this need. I think the conclusion is just that the understanding will be much more difficult, and is much further off, than any simple answer we can state to give our lives a purpose. But that's okay; otherwise it would not be so interesting, or so significant, if we ever do understand it.

These meditations under the stars were meant only to remind you, to put you in a mood of greater awareness of these questions, and to recognize the interaction with your environment that must occur to generate any kind of answers to these questions. This is important because the questions are often vaguely defined and easily lost among everyday concerns, so the atmosphere and the mood are important for understanding them. It's something like the feeling one gets in listening to a powerful piece of music: it seems there is some meaning, something wonderful going on behind it all, but the feeling is easily lost when the music stops.

Now that we have some broad, long-term ideas of the kinds of things we'd like to think about, maybe we should start asking more specific questions, looking to everyday activities that are very familiar, as a way to get started on our questions. For me, running provides a good example. I enjoy the competition and the challenge of trying to push the limits of what I can do. When I'm running in a race, it always seems that, if I have the will power, I can make myself go just a little bit faster. I always feel I have the *choice* to push a little harder, and it seems that I should be able to continue this process indefinitely. But something is wrong. I haven't yet made it to the Olympics, and I'm not optimistic about doing so! I cannot run as fast as I wish. What is the relation between my free will and the limitations on me? I can force myself to stay awake to finish a last minute project, or to run faster, or perhaps to stay alive if I'm badly injured, but only up to a point. Eventually I must sleep, must slow down, must die.

This relates closely to our earlier questions about purpose and meaning. In order for our actions to be significant, we want to be able to take at least some credit for them. I want my success on the history test to have had at least something to do with my own efforts. How could I feel proud of myself for my success if I had no choice in the matter, if I did well simply because I was lucky enough to have a good memory, or to have been told so often that I should study, that I did it without thinking? The idea of fighting against odds, of struggling to succeed, has significance only if there is something real to struggle against. We need some way for our actions to be limited by constraints, yet which still allows us to struggle to overcome the constraints.

Why is the universe such that it has the right kinds of constraints and limitations to allow this struggle? It causes pain and suffering, but also hope and glory. This is a deep mystery, and one that will likely be a focus of our thoughts. But again we seem to be at an impasse. How do we go about understanding the mystery better? How do we discover how the world came to be the way it is, and why it is that way? The last example of thinking about running gives a hint toward a good approach, which will bring us back to the perspective on science with which the chapter began. We have ended up asking very fundamental questions about free will, but because the questions were tied to a specific situation, we are not working in an abstract vacuum. We have a way to focus our thinking, an example to go back to if we start getting lost, and something specific on which to test out any new ideas we may come up with. This is the value that I think ideas from science can offer to our efforts to understand our place in the universe.

Of course, we don't have to believe that we should try to find out why the universe allows the possibility of struggle to overcome difficulties. We could just accept that it means there is some purpose we should try to carry out. Or we could give quick answers that propose to solve the mysteries, but don't really solve them on a deep level. Then we could believe that we really are pursuing the purpose set out for us, and devote all our energies to it and unleash the full power of human creativity on this goal.

In other words, we could settle for stories that explain what we don't understand in terms of our own very limited experiences so far. This is fine for a start. It seems best to begin with the familiar in order to have something to work from. In fact, that's really all we *can* do—we can only start from what we know. Generating scenarios for how the meaning we seek might be implemented in the real universe is an important part of our progress. The problem comes when these stories become enshrined as the final word on the questions they address. This may give us security, and cover up our fear of the unknown, at least temporarily. But it would be a shame, because it would cause us to stop asking questions, at a point so early in our understanding that we're only just beginning even to understand what questions to ask. By settling on final answers based on very limited knowledge, we risk stifling the sense of cosmic perspective that was beginning to open up for us. And ironically, this would also betray the very goals, the desire for meaning and purpose, which make us want answers and certainty in the first place. After all, if we're going to be passionate about wanting to do what's right, don't we owe it to ourselves to explore, as widely as possible, all the ideas about what's really right?

Fortunately, there is another option. This is the option that I hope you will develop or become aware of as you continue to read and to contemplate the role of science in your questioning process. Perhaps we cannot start out by answering such big questions as, "Where did the universe come from?" or, "What role do I play in the context of the universe?" These questions only guide our efforts, leading us toward the understanding we are after but are unsure of how to clearly define. We can, however, begin with smaller pieces, to work toward a better understanding. We can start out by asking more manageable questions about the universe as we see it around us, keeping in the back of our minds that we want to work toward the answers to our original questions.

We should keep in mind, as well, that our original questions may change as we gain a clearer understanding of this "something" that we are after. Our subsequent learning may cause us to completely change the questions we ask. This does not mean that the original question was a waste of time. Without it we would not have had a starting place, and might never have made any progress at all. It gives us a path to follow, and once we are moving along a path there is hope for progress, even if the progress leads to something other than our original destination.

With these reflections as background, we can reconsider the question that opened this chapter: What does science have to do with your personal worldview? The answer, I think, begins with another question: Do you believe some of your actions, and the outcomes produced by some actions, matter more than others? I suspect most everyone will answer "yes" to this question. We certainly behave as if we believe some actions are better or more important than others. Every time we criticize or praise someone's behavior, for example, we are asserting this belief. So then doesn't the essence of what matters in life, what drives our path through life in a positive way, boil down to figuring out what actions really matter in some deeper context, and then gaining the power to do those things? But this task is quite overwhelming, as anyone who has ever sat down and tried to "figure things out" all at once surely knows. Our thinking quickly goes in circles. We come up against contradictions, things we seemingly cannot ever know, and can end up thinking it's all hopeless and there is no way to make progress of any kind at all. My suggestion here is that science and its key insights, while far from providing all the answers, can provide a guide and a structure to our process of moving through life in a way that makes progress in understanding a possible context within which our actions matter.

Reflection and Discussion

• Self-reflection questions to help strengthen your awareness and connection to science and the world around you – As we ask ourselves what our lives mean and how we fit into things, information from science becomes valuable for shedding light on the overall framework within which we try to construct meaningful lives for ourselves.

These questions provide a way for you to develop a dialog with yourself (and others, if you wish) about the kinds of topics that matter most to you. They will help you become more actively aware of your connections to the rest of the universe, and of which insights from science might be most valuable in helping to answer your questions. By answering some of the questions you can begin to turn this into an active process of discovering connections for yourself. They are intended to trigger your thinking and connect you to ideas that will help you answer the general question:

> *What are the most important questions you have (related to understanding your place in the scheme of things) that you wish scientists would address and provide you with information about?*

For many of the questions listed below, there are no "right" or "wrong" answers. They are meant to help you become more aware of your own thinking about your world. Pick and choose questions that are interesting or helpful to you.

These questions may also be found in electronic form at www.scienceintegration.org. You can submit your answers there, to get involved in a dialog with others who are pursuing similar questions.

Awareness of your surroundings

• Have you ever seen a full moon in the middle of the day? Where was it in relation to the sun and in relation to the directions (N–S–E–W) on Earth?

• What is the nearest plant to your front door? Does it have edible, medicinal, or other uses?

• At about what time did the sun rise and set yesterday?

• What phase is the moon in now and at about what time will it rise tomorrow?

• Where is your nearest source of fresh water?

• What was the first thing you thought about when you woke up this morning?

• When do you pay the most attention to the world of

nature around you? Early in the morning when the birds are singing? At sunset? Whale watching? Hiking? Feeding pigeons? What kinds of activities most easily put you in a frame of mind where you are aware of the big picture of the world you are a part of? When do you feel most connected to processes extending beyond your immediate surroundings?

• Are there elements in our solar system that are not found on Earth?

Thinking about origins

• Where did the Earth come from?

• Where does life come from?

• Does it bother you to think that humans evolved from other species? Why or why not? If it does bother you, can you come up with a way to interpret it that might make it more acceptable?

• What are our bodies made of? Where did this material come from? Trace it as thoroughly and as far back as you can.

• Why do you think we developed the ability to think and feel? How did consciousness develop, and what purpose does it serve for the universe that produced us, that we are able to think?

• What are shooting stars? What are they made of, and what causes them?

• Do objects from space ever land on Earth? If so, how often, and how big are the objects? What are they made of? If not, why not?

Cause/effect links and interactions

• What are some of the ways your thoughts affect the world around you?

• What are some of the ways the world around you affects your thoughts?

• What positive effects do humans have on the Earth?

• What negative effects do humans have on the Earth?

• What effect does the moon have on the Earth, if any? What would be different about the Earth if the moon did not exist?

• How would your bioregion be different if the Earth's average temperature were 10° F warmer (or colder)?

• Why does the moon have craters? What caused them? Why don't we see craters like them on Earth?

• List some of the ways the sun affects you (as many as you can). How is your life connected to what goes on in the sun?

• What would happen to our solar system and to Earth in particular, if Mars suddenly disappeared?

• Do you know what happens to the trash you put out to be picked up in your neighborhood?

• Do you know what happens to materials you put out to be recycled in your neighborhood?

• Why do you think an old song or the smell of a flower

can bring back memories so instantly and vividly? How does this work?

Relating scientific explanations to everyday experience

• What does it mean if a red-tailed hawk is hanging out in your neighborhood? What else must also be there?

• Describe a wind in terms of the atomic theory of matter. Can you make this explanation fit with your direct experience of wind?

• If your chair is made of atoms that are mostly empty space, why don't you fall through it?

• Why is ash harder than cedar wood? What is different about the structure of the wood in the two cases?

Attitudes and perceptions about science

• What comes to mind when you hear the word "science?" Don't evaluate or filter your answers; just write down a few things you automatically associate with the subject.

• Summarize your previous experiences with science. Particularly helpful will be experiences not in a science class. What questions about the world have most sparked your curiosity? What experiences with science have most turned you off of the subject?

• Describe how you use the information you learn in science classes in your daily life, and how this information impacts the way you perceive the world. Are there any ideas you associate with science that you feel have significantly changed how you act, how you live?

• List some beliefs you hold (or once held) that you would classify as "superstitions." Where do you think these beliefs come from? Why do (did) you believe in them? What makes you label them as superstitions?

• Do you think it is important to understand things like why the constellations are in different places at different seasons, why cats have retractable claws, why there is mostly basalt rock around your city, or how your cells work? Does it make any difference to your life, to know such things? Does such knowledge change the way you look at your life in relation to the universe? Does this knowledge make you happy or sad, feel purposeful or obsolete?

• Do you think science takes away magic, purpose, or fun in life? If so, can you think of ways it could it be changed (either in how it's practiced or how it is taught) so it would not do this? In other words, try to identify specific things about science, either its attitude and approach to the world or specific discoveries it has made, which make you feel it has taken magic and purpose out of the world. On the other hand, if you think science adds magic, purpose, or fun to your life, try to explain why.

• Do you think you would be happier if our culture told and believed in myths about nature that made the world alive and magical and purposeful, even if these stories were "inaccurate?"

• It seems to be human nature to complain, to wish things were different than they are. So here's your chance to remedy all these problems! In the last chapter you described the universe as you actually perceive it. Now describe the properties of your "ideal universe." What kind of universe would make you feel most welcome, most at home, most certain that your life was meaningful? How would such a universe operate? What would be in it?

• Which aspects of your ideal universe (if any) do you think are contradicted by the findings or the worldview associated with science?

• Do you consider yourself to be part of nature? What's the basic difference between something that is natural and something that is artificial?

• How do you decide whether the use of certain technologies is right or wrong? Consider nuclear power, genetically engineered foods, or driving a gasoline-powered car, for example. What factors enter into your opinion about whether these technologies are good or bad?

• Is global warming bad, if it is caused by natural changes in the sun? Is global warming bad if it is caused by the increase in carbon dioxide produced by humans? Why/why not, in each case?

• Are forest fires bad? Why/why not?

• Are earthquakes bad? Why/why not?

• What fundamentally makes something good or bad? What criteria do you use to assign these categories?

• What are the most important things you've learned from science?

• Is the scientific view of the universe beautiful?

• Where do you get your beliefs about the great questions

of life: Where do I come from; what is my purpose here; etc.?

- Do you think science can provide all, some, or none of the answers to these kinds of questions?

- Where does the authority of science to describe the world dwindle, in your mind? Where do you draw boundaries between the parts of your experience where science applies, and the parts where it does not? Do these boundaries occur at questions of consciousness, morality, religion, or somewhere else, or not at all?

- Do you think science is the only, best, or worst set of tools with which to reliably interpret the world?

A wider awareness of the universe and your connections to it provides a greater variety of ways for you to find meaning in your life, as suggested by this quotation:

> *[Our] role consists of widening and broadening the visual field of [a person] so that the whole spectrum of personal meaning becomes conscious and visible to him.*
>
> —Viktor Frankl[2]

Chapter 3

Applying this Perspective in Your Own Life

Summary: This chapter provides examples of how we can approach events in our lives to feel a direct connection between our daily actions and a deep, broad context that makes these actions matter. How can we feel ourselves contributing directly to some greater context, something that is significant, with every action we take? And most importantly, how can we organize an approach that makes real, tangible progress regarding deep questions that are notoriously slippery?

A healthy consciousness is like a spider's web, and you are the spider in the centre. The centre of the web is the present moment. But the meaning of your life depends on those fine threads which stretch away to other times, other places, and the vibrations that come to you along the web...Normally, your consciousness is like a very small spider's web; its threads don't stretch very far. Other times, other places, are not very real to you...And our lives are turbulent, like living in a strong wind, so the web gets broken pretty frequently. But sometimes the wind drops, and you manage to create an enormous web. And suddenly, distant times and distant places become realities, as real as the present moment, sending their vibrations down into your mind.

— Colin Wilson[1]

A few specific examples may help illustrate what it might be like to live in a way that uses science as a core part of how you view your relationship to the world, and as an important tool in your worldview-development process. The common feature of the frames of mind built up by these examples is a sense of connection to a bigger system, which helps you feel part of something that matters. In one way or another, when we seek meaning we are seeking a way to view our individual actions as grounded in a broader context. The aim here is to provide a glimpse of how you might use a perspective informed by insights from science in order to do this in your own life. As I hope will be clear from the range of examples, you could truly do this with *anything* in your life. My examples serve only as illustrations, to point the way. It's important to emphasize also that this process will not give you immediate answers. Rather, it helps you hook into a *process* that provides an organized way to generate ideas and try out answers, in a way that is connected to your ordinary experience and concerns. It allows science to feed into your meaning-making effort, rather than leaving it compartmentalized in a detached and abstract realm. The perspective is guided by this idea: We seek to ground what we do in something deeper. But often, we don't connect our own daily lives with the universe out there, to see that there is something *real* going on, that our struggles can truly matter. The thinking process I'm trying to illustrate is really just a way to guide yourself into these connections, so that you learn to live with a concrete and immediate awareness of them.

So let's begin with something very immediate to your experience right now, something that may also seem rather mundane and ordinary. Consider how you are able to read and understand the words on this page. That seems a fairly straightforward question. But try to imagine all that must happen in order for you to become aware, by looking at a few scratches on a piece of paper (or dots on a computer screen), of what I was thinking when I made those scratches. To keep things as simple as possible, let's just ask how we are able to *see* the scratches at all. How do the patterns on the paper or screen travel the necessary foot or two through the air to get to your eyes?

With a little imagination, you could dream up many scenarios for how the information is transmitted. Tiny winged messengers might be flying back and forth to tell your eyes what is on the paper. Perhaps we are immersed in some invisible fluid that is changed by the words in a way that in turn affects our eyes and allows us to perceive the words. This could happen without the need for anything physically "traveling" between our eyes and the paper. Maybe there is in fact no physical link at all, and the information is somehow passed to you instantaneously. After all, there is no reason to believe that what happens at one place could not simply change something at another place. The words on the paper might be instantaneously transferred to my brain during the act of looking at the page. Maybe nothing else about the process can be described. We know we can at least *imagine* such a reality, for example in science fiction stories involving psychic powers or telepathy. Try to think of other possible descriptions ("models") for what's going on that enables you to see the words. As soon as you start to think about different models, and to consider which ones have the best chance of being correct, you are doing science, in the spirit of the perspective I suggested in Chapter 2. You begin to be drawn in to an awareness of the connections between yourself and the external world. And you can start to see that your mental models for how these connections operate have consequences that can be compared to your other experiences with the world, to see if the predictions of your model are consistent with reality.

Arguments over which of our imagined explanations is correct are not of much use unless we have some way to distinguish between them in the real world. If you say that light is caused by a being called Mercury flying back and forth between the paper and my eye, and I say, no, it's really Thor, but we both say that the only distinguishing characteristics are the way they carry light, then we really haven't said anything. This is not just a statement about science; it applies to any meaningful beliefs about reality. In this more general sense, people may disagree about what constitutes a test of an idea, but very few would be interested in an idea that has no noticeable consequences of any kind. We wouldn't argue about which being was responsible for light if there were really no difference in the two statements. But, for example, arguing that you should offer a sacrifice to a being of one name or the other in order to get what you want is a *real* difference, which could in principle be tested. If there really were no observable or measurable difference between one belief and the other, then no one would argue the point. I think it's important to recognize this because in discussing questions of meaning and our role in the universe, certainly ways of knowing other than science come into play. But ideas from these other ways of knowing can still work or not work, and are still subject to constraints imposed by the external world. We're not free to believe anything we like just because we're no longer purely in the realm of science.

In any case, somehow we are able to create in our minds various ideas about how reality is. Light does certain things as it travels through space, and we're able to make up an idea that exists only in our thoughts, work through what would happen in a certain situation according to that idea, and then compare it to what really happens. Some of these ideas are wrong, and some may have varying degrees of truth to them. Those that agree with reality need not be dull, as scientific explanations sometimes seem to be. They can still have human meaning in them, while remaining consistent with what we learn about how the external world works, and with what we know about the specific constraints and limitations it imposes upon us.

Fortunately, for the models or theories I have proposed in an effort to explain how we see the words in this chapter, we *do* have ways to distinguish among them. We can deduce the implications of a particular explanation, as logical consequences of our model. Here again, I emphasize that the use of logic is not unique to science at all. The process of drawing logical conclusions from an idea is crucial to any attempt to know the world around us in any meaningful way. In fact it is intricately tied in with what it means to be able to know something. We're always piecing things together by linking them with logical steps. I may recall that I saw a friend on the way to work this morning. As we were discussing plans to meet for lunch, I noticed that he had on a blue baseball cap. Later that day as I walked to lunch, I spotted an unrecognizable figure standing in front of the cafeteria. All that I could tell about the person was that he was wearing a blue cap. From all this information, I know with some confidence that it is my friend waiting for me (I am five minutes late for our lunch meeting). The point is simply that the use of logic to draw conclusions in science is nothing fundamentally new or magical. It is simply a generalization of the process we normally go through to know something about the world. Let's continue with our investigation of the words on the page…

In order to test our ideas, we might be able to measure the travel time for a light signal, in which case the instantaneous travel theory could not be correct. This turns out to be the case. If you are 300,000 km (186,000 miles) away from a friend who turns on a flashlight, you have to wait an entire second before the light will reach you. As you might imagine, this travel time was not so easy to measure, and for a long time it seemed that light did travel instantaneously. It's worthwhile to think about ways you might try to measure it, and what would be the fastest speed measurable with each method.

Given this experimental fact, it seems a reasonable starting point to assume that there is some sort of physical messenger to carry the information. Following this line of reasoning, how might we learn more about this messenger?

If you're reading this at night by lamplight, you can easily discover (by turning the lamp off) that the lamp plays a very important role in your ability to read these words. However, if you're sitting near a window in the daytime, the lamp is of little importance. The difference seems to be that the sun is shining.

These statements may seem too obvious to be worth mentioning. You may be beginning to wonder about the quality of my science training, that I am so fascinated at discovering such obvious connections. Of course we know that you can't see in the dark! But in fact we have observed something quite remarkable that we normally take for granted. When we say we need light in order to see, we haven't really explained anything, because we don't even know what light is. At this point we could just as well have named it Mercury, and have said just as much. Light, or Mercury, are only names. By themselves they tell us nothing. But, by comparing our ability to read at night and during the day, we've discovered something much more remarkable and useful. Whatever it is that we do need in order to see (go ahead and call it light if you'd like!), it seems that the distant sun and the little lamp on my desk both accomplish the *same* thing. In fact, if we're looking for explanations, it's very tempting to suggest that maybe they *are* the same thing. Maybe the light that comes from the sun is the same phenomenon as the light from the lamp. This suggests a whole series of additional experiments we could carry out to test this wonderful new insight.

If it turns out to be true, this would be a very amazing statement about the world, far beyond its innocent beginning as an interest in knowing how we see the words on a page. The sun and the light bulb are totally different things, in different places. Yet both produce this substance, light, which is needed for us to see anything. The insight that nature has general principles that are *consistent*, the same everywhere we look, is a great one, which grows deeper the more we think about it. It is a key observation about our cosmic "stage" that we will want to file away and incorporate into our efforts to understand the "play" we are a part of. Science provides a process and a way of thinking that allows us to come to these kinds of realizations. This is why some kind of understanding of it is valuable to poets and philosophers and anyone who wants to understand our place in the world.

The existence of consistent principles for how things work is a remarkable fact about our world as we understand it so far. It is also a very fortunate circumstance, since this fact seems necessary if we are to have hopes of really understanding nature. Many similar discoveries await us as we continue to explore, some of them even deeper and more profound, such as relativity, cosmology, and quantum physics. No religion or general philosophy of the world could hope to be "correct" in the deepest sense, if it does not take into account some of these insights.

One other point worth noting here is that we can never be certain we have identified "the cause" of our ability to see, or that the cause is the same in both cases. We haven't *proven* that the light from the sun is the same phenomenon as the light from the lamp. Maybe it's very different stuff that just happens to act the same in certain circumstances. It's worth being careful about this, but in fact we never completely prove anything. We just make reasonable guesses that enable us to carry on with life. We don't know for certain that the sun will rise tomorrow; we haven't proven it, but it's a pretty reasonable belief. If we want to press it too far, we're never really certain about anything, and could never get anywhere. So, we might as well go on and keep trying to draw conclusions to live by, but simply proceed with caution.

So, now we've established (we think) that this stuff that comes from the sun and from lamps is necessary for you to read these words. But we're still not anywhere near being able to explain what's happening. (Knowing that an airplane is necessary in order for you to fly across the ocean, or a rocket to travel to the moon, doesn't tell you how to do either of these things!) We need to discover how this light behaves and interacts with things, if we hope to find out how it helps us see. What we're claiming so far is that there is something, light, that carries information through space from an object to our eyes. The sensation we call seeing the words is an interaction of this light with our eyes. Furthermore, the light necessary for this to happen is not always present everywhere. There are certain conditions which must be met in order for it to be there. Two things we have discovered that are sources of light are a table lamp and the sun. Based on what we've talked about so far, this is all that we can confidently say. In order to say more, and to make predictions in new situations, we need to study more about how light interacts with other things. Specifically, that means, "How does what we see change when we put different things in the system?"

To do this, we start by doing some experiments that you might want to try for yourself. The first is one situation in which you're lucky if you're nearsighted. Get a small piece of aluminum foil and put a tiny pinhole in the center of it. Then take off your glasses or remove your contacts and find a street sign or something that is well lit, but far enough away that you cannot read it without your glasses. Now look at the sign through the pinhole, and you'll probably be able to read the sign, if there's enough light and it's not too far away. Try to understand why the pinhole enables you to see clearly what you could not before. (Incidentally, a pinhole camera works, and has tremendous depth of field, due to the same effect. So you can explore this property of light with a pinhole camera even if you have perfect vision.) In particular, what does this experiment tell you about the properties that light must have? Could it be made up of individual particles that stream from the object you're looking at to your eye? If so, what must be happening to the particles as they pass through the pinhole? What does it even mean, in terms of the properties of the light, for you to be able to "see something clearly?" That is, what happens to the light that's different for a clear image as opposed to a blurry image? Can you describe these two situations in terms of your mental model of light?

The details of what you come up with here are not so important, but I hope these examples are starting to get you into the flow of this way of thinking. In some amazing way, it provides a process for coming to deep insights about how the world works. It's through this process that I'm suggesting science can have an important role in your attempt to build a meaningful life. It doesn't give us immediate, definite answers to our questions about what our lives mean. But it gives us a way to make real progress, to collect together insights in such a way that we know we are moving forward, getting closer to understanding, and not just spinning our wheels.

It's worth noticing that much of the science that can contribute to your daily sense of perspective is right in front of you, all around you, in the "ordinary" world you live in. There is a tendency for us to think that the exotic, truly mind-stretching ideas at the edges of our knowledge of science, which are far beyond common sense reasoning, are the only places to find the deep mysteries and clues to the meaning of life. Perhaps this is because we know we don't have the answers from what we see immediately, so we expect that the edges of the unknown may hold those answers. Also, it's easy to forget that the reasoning going on in these exotic realms of science is just an extension, the result of step by step buildup from immediate experience—it's not fundamentally different. In any case I think it's valuable to remember that no mystery we could find anywhere in the universe is deeper than the simple fact that anything at all exists, in this moment.

Let's develop similar connections with an example from Chapter 1. Again, our goal is to illustrate a thinking process that grounds the experience or issue in a deeper awareness of the big picture and of the reality on which the idea is based. I invite you to take these examples as an illustration of the process you can use to change your awareness and relationship to whatever matters most to you, in your world of daily experience.

Like many people, you may think that recycling is important and virtuous. But why do you believe this? (Or why not, if you don't like recycling?) Perhaps you recycle because you value protecting the Earth. But on further probing, you can see that you must have some context in order for this belief to make any reasonable sense, or even to be clear on what you mean by the belief. Protecting the Earth in what way, for what purpose? What do you think is going to happen in the future, that matters, for which the Earth needs to be in a certain state, a state that will be achieved if we recycle but may be destroyed if we do not? Our beliefs on this matter are clearly context dependent. They depend on what we think is going on overall, what our goals are, how we fit in as part of the universe. As an extreme example, recycling is probably not terribly important to anyone who believes the universe will come to an end long before we would ever use up readily available raw materials at current rates. To say we need to preserve things for the future is to implicitly assume that there *is* a future – and a significant future goal whose success depends on our recycling. Do you want to protect the Earth for future generations of people? How many generations? Or maybe you're not so concerned with humans in particular, but instead have an idea that there is a natural order of things we should not disrupt too much. This is fine, too, but it's certainly an entire worldview of its own, that deserves to be articulated and clarified if we're using it as the basis of our actions. It brings into play a network of ideas about how we came to be, what makes us different from other organisms in nature, and what gives us special power to mess things up. And the choices we would make in the context of this set of beliefs are in many cases different from those we would make if we were concerned purely about the consequences of disrupting the life support system the Earth provides for us as humans.

Of course, we don't usually think consciously about these things every time we take our glass jars to the recycling bin, but the beliefs are present as a foundation for that action. An effective way to bring these background assumptions to the surface is to ask yourself questions about what new knowledge you could gain that might change your beliefs (and actions). How would your feelings about protecting the Earth be affected if you knew more about the process (which took billions of years) that occurred to produce the conditions on Earth that are so well suited to life? How would knowledge about the conditions on nearby planets affect your decision? What if we learned of a new, vast source of energy that made fossil fuels unimportant? What if we learned that our solar system would be engulfed in a supernova in a thousand years? Try asking questions like this of yourself, and see how your views would change in light of hypothetical new information you could learn. If you're clearly aware of what new information would be required to *change* your opinion, then it's much easier to see what the real *basis* of that opinion is.

For the sake of continuing the illustration, let's assume you do think it's important to preserve natural conditions on the Earth. Even if we agree on the *aim*, another important question arises regarding the *method*. Is recycling the best way to go about preserving the Earth? To really answer this question, to make sure that your actions will have the *consequences* you intend, you need to examine even more information about how the world actually is. You need to have (or trust that someone else has) a basic understanding of how matter is transformed from one material to another. You need to know how much of a material is readily available. If it is a renewable material, you'll want to know the time-scale of the renewal process. Can it be replenished in one year, a hundred years, or a billion years? You need to be aware that the industrial processes that are required to recycle materials (collection, sorting, cleaning, etc.) have effects on the environment themselves. In some circumstances you may find that it's better for the overall environment to throw away some materials than to recycle them. It's not always a clear-cut answer. This is getting complicated!

To make the example even more concrete and immediate, let's look at a specific, simple choice you might be faced with—just one of the many little choices you face as you work your way through a day. Consider an empty glass juice bottle you hold in your hand after drinking the contents. What are you going to do with this bottle? Suppose you have just two immediate choices: You could toss it in a nearby garbage can, or you could put it in a glass-recycling bin. Each of those actions will set in motion a whole series of other actions by various people and machines, and will have an impact of some kind on the condition of the Earth (pollution, energy use, natural resources available, etc.). Which option should you choose? Most of us would probably choose the recycle bin (especially when under the spotlight of public scrutiny), because we make certain assumptions about the consequences ultimately produced by that action. But most of the time our awareness of how our action connects to that outcome we desire is very tenuous and very vague; we just have some vague idea that "recycling is 'better' for the Earth," without being sure what we mean by better, for whom or what things on Earth it would be better, or where we learned this belief.

I should emphasize clearly that I'm *not* trying to get you to oppose recycling. It probably is the best thing, for many materials, by some reasonable definition of what we mean by best. What I am trying to get you to notice is that the answer is not obvious; it is *information dependent.* Hence it's crucial that we know the science at some level, i.e. how the world will work in response to our actions. Depending on the type of material, how it is produced, and how the recycling process for that material works (including how the recyclables are sorted and collected, by curbside pickup or some other means), recycling may have more or less negative impact on the environment than throwing the stuff away, waiting a very long time for nature to recycle it, and in the meantime going out and collecting more from nature. We cannot legitimately make these kinds of decisions in the absence of information about how things will work. Without such knowledge, our moral code may be leading us in exactly the opposite of the desired direction.

It seems to me that the basis for all such views we hold can be traced to some core unifying principles about how the world works, which the process of science helps us to gradually build up and identify. For example, our views about the essential moral goodness of resource conservation, recycling, and related environmental concerns ultimately owe the foundation of their validity to a law of physics known as the second law of thermodynamics. This is the fundamental property of nature on which these values, expressed through our specific attitudes about waste and conservation in specific situations, are ultimately supported and made valid. It reflects our fundamental inability to sort things back into useful form without creating even more disorder in the process. We may not be aware of this link at all, but if we were to change this law, the universe would be different and many of our views would change. In a universe with different rules, a different set of values on these topics would win out.

Imagine, for example, a universe in which concentrated energy was increasing at such a rate that we had to convert it to other forms quickly ("use it up") to reduce the risk of explosions, or in which trees appeared from nothing, so fast that we had to keep burning them to avoid being overrun. Obviously, our world does not work this way. But we would have different views about what actions are best, if the world did work this way. So it provides a simple example in which a moral value ultimately depends on fundamental laws of the universe. If we change the rules, we change what is good. So, if we change our *understanding* of the rules, we also change our understanding of what is good, and hence we change our worldview and our actions. Once again, our hidden assumptions and beliefs are at work behind such a simple action as what we do with an empty glass bottle.

You can draw out these assumptions about the world beginning with nearly any preference or decision you encounter in your daily life. You'll find that all of them contain some assumptions rooted in your beliefs about how the world works, beliefs that can be influenced by new insights revealed by science. Try asking these kinds of questions about your own preferences.

Let's summarize the approach we're trying to develop through these examples: We want to approach each of our significant beliefs and actions by tracing them back far enough to see how they are grounded in beliefs about what the universe is like and how it basically works. We then want to actively try out belief scenarios on those terms. Given that the world works this way, does my belief or action make sense? What would I modify if new information about the world became available? We want to set up a channel for learning about the world in a way that connects directly to our personal worldviews. We'd like to live in a state of awareness from which we are conscious of this link, so there is an ongoing interaction between our beliefs and our awareness of the world.

I want to be able to see that I recycle because it is necessary for certain objectives, given certain beliefs about how the universe works. Then I'll feel very confident about my recycling activities, confident that my actions are serving the objectives I really believe in. And, I'll be in tune with what I currently believe, so I'll be prepared to notice new information about the world that challenges any part of my current worldview. This puts me in a better position to modify my beliefs and actions, if necessary, to be consistent with new information.

As I acquire new information and perhaps modify my beliefs, I also want to guard against losing the immediate, tangible sense of connection to the world that makes it feel wonderful and magical rather than ordinary and dull.

> *When I heard the learn'd astronomer, / When the proofs, the figures, were 'ranged in columns before me / When I was shown the charts and diagrams, to add, divide, and measure them / When I sitting heard the astronomer where he lectured with much applause in the lecture-room, / How soon unaccountable I became tired and sick / Till rising and gliding out I wander'd off by myself, / in the mystical moist night-air, and from time to time / Look'd up in perfect silence at the stars.*
>
> —Walt Whitman[2]

This passage from Whitman expresses the sentiment that as we learn more about the world through science, it seems somehow less wonderful, less enchanting, less a place we feel at home in. But our understanding of the universe is still in its infancy. How carefully considered is the intuition that the universe which science is helping us to uncover might not have a very definite and clear context within which our actions, our beliefs, hopes, dreams, and feelings, might fit? This is the pathway we'd like to open up, and it's an enterprise we can all participate in, as the examples in this chapter have tried to demonstrate.

Reflection and Discussion

As you've learned by now, I can't end a chapter without giving you something to do. So here are some questions to help you articulate your views about science and its relationship to your search for personal meaning. A few should look familiar from Chapter 2, but here they are asked from a slightly different perspective.

- How would you define science? What characterizes it as a method of inquiry distinct from other ways of knowing about your world?

- Is science (as you've defined it) a necessary part of your life? If so, describe how you use science as part of your daily life. If not, do you think we'd be better off or happier without science?

- What qualities or properties do you think the universe must have in order for you to feel that you have a place in it, and that your life is meaningful? (In other words, what are the key features or properties of your "ideal universe"?)

- In what ways has science influenced your sense of what kind of universe we actually live in? (In other words, what do you perceive as the key features or properties of the universe science tells us we live in?)

- What beliefs do you currently hold, that are both important to you and that you think might be considered "unscientific?" (i.e. beliefs you think would be criticized by scientists, or that you know contradict accepted scientific evidence, or your perception of the scientific view of the world.)

• Do you think science could be interpreted or applied in a different way that you might find more satisfying? If there are aspects of the world described by science that you do not like, try to evaluate whether it is the *facts* or the *interpretations* that you don't like.

• What new facts could be discovered, that would strike you as evidence that this is a meaningful universe, of the sort you'd like to live in? (For example, proof that life exists on other planets, or evidence that telepathy or astrology really works, or evidence of "miracles" of a certain sort, or whatever.) The aim of this question is to bring to the surface the essence of the kinds of things you're looking for. Then you can search more actively for them, using the perspective science might help you gain, rather than working against information we've learned from science.

Chapter 4

Why Isn't Science Already More Integrated into Our Lives?

Summary: This chapter voices some of the common associations with science that can get in the way of using it to help clarify our sense of perspective and meaning. The aim is to get these associations out in the open and addressed, to reduce their power to interfere with our progress.

Science! true daughter of Old Time though art! Who alterest all things with thy peering eyes. Why preyest thou thus upon the poet's heart, Vulture, whose wings are dull realities?

— Edgar Allen Poe[1]

In thinking about how to incorporate insights from science into our personal, meaningful view of the world, we're not starting from a neutral position. We are not dealing with science as just a method for figuring out which of our ideas work within the external reality we experience. Instead, we are dealing with the culture of those who carry out scientific research, with the interaction between this culture and the rest of society, with the way the methods of science are perceived and the way its insights have filtered out into broader culture and society. Science has acquired a reputation for dehumanizing the world, leaving us in some way stranded and alienated in a universe for which our existence seems irrelevant. This view of science as a cold, dry, and dehumanizing subject that takes meaning and fun out of life is widespread enough that we need to address this state of affairs head on. By probing the origins of this view, we can try to understand what causes it and perhaps see these causes as a real part of the history, culture, and practice of science, but not inherently necessary to science itself. By understanding this, we can overcome some of our resistance and free ourselves to get the full benefit of science as an integral part of making our lives meaningful.

So let's begin an exploration of science and our views of it, with an eye toward pinpointing the aspects that we may find alienating. The subject we call science serves many purposes and means different things to different people. The part of science we're generally most aware of is its application to technology. Through science we learn the rules that describe how the natural world operates. This knowledge gives us the power to control parts of nature, to arrange things so that actual events match more closely to our wishes. In today's world this ability to make reality conform to our wishes has expanded rapidly to influence nearly every aspect of our lives. The desire to feel cool on a hot summer day can be met by turning on an air conditioner. The wish to exchange ideas with nearly anyone at almost any time can be carried out by telephone or the Internet. The wish to be physically present at nearly any location on Earth can be made possible by some combination of airplane, train, helicopter, boat, and automobile (along with walking or other human-powered transportation).

These and many other technological marvels that pervade our lives seem to be the most obvious and significant impact that science has had on our society. But surely it is not directly this power to control nature that gives rise to the sentiments about science expressed in the quotation from Poe, or the general sense that science is alienating and dehumanizing. If science were only about making the world conform more closely to our wishes, to cure diseases and free us of material needs, then we'd have no reason to complain that it had taken away the magic, fun, or meaning in the world. It would simply have made us healthier and freer to experience the wonders we do find around us.

I think the common feeling that science is dry and somehow "dehumanizing" arises from the view of the world science seems to require, in order to have such great success in controlling that world. Science is not just a neutral listing of recipes for getting nature to do our bidding. It carries along with it a view of how nature is arranged and how it works, a view that necessarily impacts our sense of how we fit into the world around us. It seems to force itself into our personal worldviews in a way we're not always happy about.

As we discussed in Chapter 1, we all carry around some vague conception of what our lives mean, of how we fit into the scheme of things. These ideas are based on a wide variety of influences, from individual experiences we've had, to religious beliefs and social customs (which are themselves the result of the collective experiences and thinking of those before us). It's difficult to pin down all the sources from which these conceptions that form our personal worldviews arise, but at least two things are clear about them: First, anyone's worldview will give significance to human life (and that person's own life in particular) in some way. Second, this worldview will be at least partially influenced by a person's experiences with how the world works.

So let's probe more deeply into the view of the world the success of science seems to require. What kind of map of the world does it give us, and how does this relate to the kinds of maps (worldviews) we construct for ourselves to get through daily life? In asking this question, I think we can begin to see the problem. Science has achieved its great success by describing a universe that operates essentially *independently* of our concerns. The laws describing how a particular medication will heal an infection are specified at a level of description that knows nothing of our wishes or thoughts or anything we care about. The success of the medication depends on "objective" things like its chemical composition, the type of organism responsible for the infection, or the temperature at which the medication was stored. Success does not depend directly on whether the person receiving the medication is kind or cruel, what religion she believes in, nor, in fact, anything about the person's moral character or thoughts. Yet this description of nature, which assumes that nature operates independently of such personal, human concerns, works so well. There must be a great deal of truth to it. It must describe the basis on which the universe really operates. This is where the real problem sets in. The great success of this view of nature seems to compel us to accept it as the whole truth, but we're not so sure we want nature to be that way. We don't want to leave the concerns that form all that makes our lives meaningful totally out of the picture. Those who share Poe's sentiment do not complain because science tells them they need not die of polio. They complain because it tells them that Earth is but a small, "accidental" speck in a universe incomprehensibly vast. They complain because it tells them (apparently) that all of their hopes and dreams and feelings and accomplishments are somehow illusions, entirely controlled by impersonal laws of physics describing the motions of the particles we are composed of. They complain because it tells them, in the words of Bertrand Russell:

> *That Man is the product of causes which had no prevision of the end they were achieving; that his origin, his growth, his hopes and fears, his loves and beliefs, are but the outcome of accidental collocations of atoms; that all the labours of the ages, all the*

> *devotion, all the inspiration, all the noonday brightness of human genius, are destined to extinction in the vast death of the solar system...all these things, if not quite beyond dispute, are yet so nearly certain, that no philosophy which rejects them can hope to stand.*[2]

This point of view is in harsh contrast to the internal feelings most of us respond to, feelings that seem much more real and familiar than the strange world of science with which many of us have little direct experience. Our internal feelings give us a different message about the character of the world we live in, a message that is expressed by William James:

> *If this life be not a real fight, in which something is eternally gained for the universe by success, it is no better than a game of private theatricals...But it feels like a real fight,—as if there were something really wild in the universe which we, with all our idealities and faithfulnesses, are needed to redeem...*[3]

There is a serious danger produced by the view of science captured by Russell. We have close, daily, personal contact with the reality expressed by James. Yet we are relatively unfamiliar with the science upon which the view expressed by Russell is based. We may quite rightly, based on strong personal experience, be unwilling to give up the belief that our lives are meaningful and significant, that we as individuals really matter in some fundamental way. But we're quite willing to give up the worldview of science, with which we may have very little direct, personal experience. So, we might think that the best alternative is to give up, to a greater or lesser extent, the view of the world provided by science. The danger is that we may become convinced that if we accept information from science, we cannot accept the meaning we somehow know is real. Perhaps this is what makes us inclined to believe in "miracles" or "supernatural" events. We are desperately seeking holes in the worldview we associate with science, gaps that would allow us to verify that there is still some wonder and magic in the world, not just "dull science."

In terms of one analogy we've been using, we might say the essence of the problem is that science discourages putting ourselves clearly on the maps it produces. Yet to be fully useful to us, the maps must have us clearly represented on them. This is obvious to anyone who has ever had the misfortune of being lost, and finding a map without a "you are here" marker. Unless you can find a way to deduce your location on the map, it will not help you get to where you want to be. I think one of the key reasons we can feel alienated by science is that it describes a world that apparently leaves out and makes no connection to our concerns and experience.

I operate entirely in terms of the choices I am confronted with. What use is an abstract system that makes no connection to these choices I face and doesn't tie in to the set of concepts that makes my life uniquely my life? In order to overcome the sense of alienation, we need to express the important concepts of science in terms which link up with the concepts we operate with in everyday life. Only then can they really mean anything to us or do anything for us. In many ways the approach of science has tended to leave us with a sense that it denies the reality of our network of concepts within which we live, implying that only the abstract framework of science is real. But to take this view is to forget the origins of the abstraction that led to the scientific framework in the first place. Ultimately our everyday concept network, which includes our feelings and dreams and decisions, is all reality can be to us. We can certainly modify and adapt this framework on the basis of new information, including much valuable information from science. But any conceptual framework that establishes *no* link to this network is by definition useless in our daily life. Such an abstract framework leaves us with no sense of how we connect its map of the world and how it can guide us in the specific concerns we have to face as we go through life. It may be a grand and glorious map, but as long as it remains detached from us, we can do nothing with it and will never feel at home with it.

A useful map must provide a direct guide of some sort, for the immediate issues in our lives. I want to feel I have accomplished something significant, want to feel I have struggled to do what matters, have fought a real fight and done something to "redeem" the universe. Where in the map provided by science do I find a place to see that happening? I don't, and that's a problem.

> *On the maps provided by science, we find everything except ourselves.*
>
> — Bryan Appleyard[4]

This leads us to the next question: Is it necessary to view science in this way? Of course it seems important that we be honest and accept reality. If it *is* necessary to take this "meaningless" view of ourselves, then we should accept it and learn to live with it. But it's rather presumptuous of us with our current knowledge, to claim that this view is necessary. We don't really even understand what we're after in hoping our lives have meaning, or in saying that we want our actions to matter in some way. How can we begin to claim that our current knowledge shows conclusively that there is no possible basis on which our actions and choices can matter? Isn't it more reasonable to think that our feelings of wanting meaning *are* based on something, and to search and question and try to understand what that basis might be?

> *What is demanded...is not...to endure the meaninglessness of life, but rather to bear [our] incapacity to grasp its unconditional meaningfulness...*
>
> —Viktor Frankl[5]

Practically speaking, we simply want to have some reasonable basis for honestly thinking it matters what we decide to do at each moment. We don't necessarily need any *particular* foundation for this belief, but we need *some* foundation for thinking it might actually make a difference that we try to make good choices. I think part of what has happened is that many of the findings of science have poked holes in some of the tentative, temporary descriptions or implementations of the meaning we seek. This is difficult to deal with. Our worldviews define who we think we are, and we change them significantly only with great struggle. So in licking our wounds from this, it's very easy to go overboard and view science as destroying the very possibility of meaning to our lives.

There are of course many details behind how we arrived at our current state, our current culture of science and its interaction with society. Much has been studied about this history, and in Chapter 6 this is one of the areas I will encourage you to explore in more depth. But very roughly and vastly oversimplified, we are in the midst of a process that must go something like this:

We start out trying to understand and organize the rules describing the constraints that operate on us, within which we try to do things. As we advance in our understanding, the patterns become clear enough to codify these rules, to see how they describe much of what goes on in nature. But then we can become so enamored of these rules that we try to insist that the universe literally *is* those rules, and nothing more. It's easy to be so impressed by the success of a set of rules we have learned, that we're tempted to label as an "illusion" anything that does not fall within the realm of these rules. This may be what alienates us and makes us feel left out of the maps provided by science.

In case you haven't guessed by now, I think the feeling expressed in the quotations from Poe and from Russell is unnecessary, even if you accept what is now known about the universe through science (which of course is much more than was known by either of these two thinkers). In fact, I hope I've started to convince you of just the opposite, that the facts and insights of science form a crucial guide and framework to order our efforts in finding out what kind of meaning there can actually be in the universe. The reason science is so important, the reason that no one can afford to ignore it, is that there probably *is* a meaningful role for us in the universe to be uncovered. This makes it important to get it "right" to whatever extent that is possible. The multitude of contradictory beliefs in the world clearly demonstrates that at least some of our ideas can also be wrong. Science can be helpful both as a guide in the process of understanding our role in things, and as protection against latching permanently onto beliefs that are inconsistent with external evidence.

> *For sanity's sake, societies must now evolve belief systems to incorporate the scientific dimension. But precious little constructive help seems to be forthcoming from the scientific community.*
>
> —D.A. Rees[6]

Reflection and Discussion

- Review the definition of science you came up with at the end of Chapter 3. Keeping that definition in mind, list what you think are the *positive* impacts science has had on the world and on your life. Then list the *negative* impacts science has had on the world and on your life. Do this without filtering, just recording your positive and negative impressions related to science.

- For each item on your lists (above), go back and think about whether your impression is justified. Is this impression a fundamental part of science, necessarily connected to what science is and how it works? Or is it something that could be separated, something that tends to go along with science but would not have to.

Chapter 5

Surveying Our Cosmic Context

Summary: This chapter is a subjective sampling of insights from science that are probably important clues to the nature of the grand context for our lives. It's intended as an overview and illustration of the types of insights you might wish to learn more about in your effort to build your perspective and your personal worldview. A much more detailed survey is given by Duncan and Tyler, *Your Cosmic Context: An Introduction to Modern Cosmology*, listed in the bibliography.

What is required…is not a detailed understanding of the content of each science, but rather a kind of synthesis of many different strands from many different sciences… When this synthesis is achieved, it could well bring the excitement of scientific discovery to many who have remained unmoved by the detailed accomplishments of the individual sciences.

—Gerald Feinberg[1]

As we consider adding insights from science to our worldview-building process, we have a rather daunting task before us. Where should we start? Science has progressed rapidly during the past three centuries, so the volume of information now available is overwhelming. Which insights capture essential features of the universe we live in, that we ought to be aware of and may want to incorporate into our worldviews? Which facts can, out of necessity, be safely ignored for this purpose? Details such as the number of grains of sand on a particular beach are surely not essential, while broad concepts such as the evolution of life, the basic framework of the history of our universe, and some of the key universal constraints within which everything seems to operate probably do express fundamental principles that are important to have in mind. And how do we extract the information we need, translating it from the language of science to the language of worldviews, into a form that tells us something directly about our place and our perspective?

This chapter offers a framework to help you orient yourself within the universe as discovered by science. It is a subjective framework, certainly biased by my experience and preferences. It should provide a starting point for your own investigations, triggering your thinking and questioning about the kinds of scientific insights that might be important for you. While certainly not a complete survey of science, it's a sample of how one can look at scientific information and insights through a particular filter, for this specific purpose: What does it tell us about the context from which we derive meaning in our lives? What does it tell us about how we got here and what we might be doing here? We're looking for information that helps us gain a better overall perspective from which to construct a meaningful worldview. This is by its nature a personal, subjective process. Yet, we share a common framework, a common set of constraints within which we all must operate in seeking our individual meaning. We don't even need to go beyond well-established science in order to sketch some important elements of this framework. Much familiar science is not really incorporated into how we see ourselves, how we relate to the universe on an everyday basis. Consider whether you live with a full, integrated awareness of even the basic science you *already know* well. Do you truly feel yourself living on a planet spinning nearly a thousand miles per hour at its surface, moving around the sun at a speed of over a million miles a day, your life part of a vast web of processes spanning billions of years?

Science is built up with facts, as a house is with stones. But a collection of facts is no more a science than a heap of stones is a house.

—Henri Poincaré[2]

I'll begin with an attempt to assimilate a few of these insights into an overall narrative view, and then follow up later in the chapter with a more detailed list to highlight some specific insights. In both cases, I focus on concepts that seem important to me in gaining a broad sense of perspective about our place in the universe.

When I step back to reflect on the implications of all the observations that have so far been made of the universe, I realize that we can say some pretty significant things. The first theme is the experimental, trial-and-error nature of all kinds of processes. This concept seems to permeate our experience with reality on all levels: from the formation of large scale structures of matter in the universe, to the development of life, and even our own feelings about what we know and the social institutions we try out in attempts to get society to work as we'd like it to.

On the cosmic scale, the universe has gone through a long process (billions of years), one result of which was to produce conditions that allow conscious life like us to exist. Every part of the universe seems to play an essential role in this process. For example, conscious life apparently arises as a property of very complex arrangements of matter, arrangements that simply could not have existed in the early years of the universe, when it was a soup of mostly light and hydrogen (the simplest atom). Life as we know it requires carbon, oxygen, and various other so-called heavy elements. And we have learned that before the first stars formed, the universe had never seen any of these heavy elements. They did not exist—stars were necessary for their manufacture. The process of forming the first galaxies and stars under the influence of gravity probably took up at least the first few hundred million years or so of the known universe's existence. Once stars were formed, they began processing hydrogen into heavier elements through nuclear fusion. Some of these stars died in massive explosions or quieter ejections of their outer layers, in either case spraying these heavier elements back into the space between stars. Later stars (like our sun) then formed from this processed material, which contained the heavier elements needed for life. Some of these stars also formed with planets around them, composed of the same preprocessed material. On at least one such planet, conditions were right for life to form. This long evolutionary process has resulted (so far) in a new and wonderful form of matter that is able to be conscious of its own existence and to have feelings and to make choices and to wonder what it ought to choose to do.

This kind of evolutionary process seems also to permeate our more immediate lives. Things change. There is evolution in the universe on all scales—cosmic and biological evolution and the much more familiar social evolution. We are guided by a desire to improve things, to "modernize" social ideas as we develop greater tolerance of differing ideas, seek to create a society that is more fair to more people, learn to improve the principles of government, and so on. In our scientific endeavors, we gradually come to a clearer and clearer understanding of the principles on which nature seems to work, modifying and refining our old theories to better match with what we see in reality.

The need for this learning process is a deep mystery of existence. Why are we able to make mistakes, in which the models we construct of reality turn out to be inconsistent with that reality? Given that we are part of this universe, our bodies governed by the rules of the universe, why do we not have a direct link to an understanding of those rules, embedded within us? Why do people exist as separate, individual consciousnesses who can propose alternative models and ways of looking at the world, and can argue and disagree and hopefully, gradually, come to an understanding of how the world actually works? Why is the history of the universe one of evolution, in which the kinds of structures that exist have evolved and changed in dramatic ways, as if they don't really know the "optimum" way to do things until they find it?

In generating an overall context for understanding what we're doing, this seems a very important theme to appreciate. How can we put these observations together into part of a story of what is fundamentally going on in the universe of which we are a part, a story that will give a tentative answer to the deeply felt question of what it all means? This piece of the story might give us a framework for building an interpretation of many common experiences. For example, mental states and feelings we consider wrong or distracting and struggle to fight against now could be feelings that once served a purpose that has been outgrown. Some expressions of emotions like prejudice, anger, or guilt might be examples. These feelings may have played important roles at one time in helping our ancestors to avoid harmful situations. Now the remnants of these feelings may often do more harm than good, and by developing our understanding of the evolutionary processes of which these are a small part, we might be in a better position to guide our thoughts in more productive ways.

Continuing on, the next major property of the universe that strikes me as important is the overall *universality* of the rules by which it operates. Even though things look very different in different places, there are a few basic principles that seem to apply to everything, everywhere. For example, the law of gravity seems to be universal: A certain amount of mass pulls on another mass with a certain force, whether the material is on Earth, on the moon, or in the Andromeda Galaxy. The same basic elements (hydrogen, helium, oxygen, etc.) are found everywhere we have been able to look (though they may be in different amounts and combinations). The speed of light through empty space is the same here as it is anywhere else. Even the laws governing such a subtle phenomenon as the passage of time display this universality. Whatever it is in my local region of space and time that tells the crystal in my watch to vibrate at a certain rate, tells another watch on the other side of the Earth (or of the galaxy, for that matter) to vibrate in the same way. And if another observer across the galaxy happens to be moving relative to me, then his watch ticks at a different rate, but in a well-defined way determined by Einstein's theory of relativity. The *laws* governing the passage of time are the same.

Though we take it for granted, this universality is really quite surprising. Why couldn't far away places have totally different laws of nature than those we observe on Earth? This fact must be an important clue to something, so it's a general property of the universe to note.

Another element of the perspective I gain from science is that everything is so much bigger and emptier than it seems when I'm trapped here on Earth. For example, where I live on the surface of Earth seems a long way from where I work, and there are many interesting things to notice during the ride between them. But even traveling at a speed that would cut my train ride to a about a thousandth of a second, it would take me a couple of million years to get to another galaxy, without encountering much else during most of my journey.

As I continue my imagined trip out through the sea of galaxies, it's easy to be struck by an overwhelming feeling of insignificance, in light of how enormous everything is. Although I can imagine I might see other civilizations on planets scattered throughout the universe, they must be incredibly far apart and unable to communicate with each other in any very meaningful way. Energetic galaxies shoot out jets of material that could engulf our entire *galaxy*, completely oblivious to the existence of our tiny planet or the even tinier city of Portland. But if I close my eyes to fight off the feeling of being lost at sea, and put together everything I know about the universe, my feeling changes: I don't think my life is so insignificant after all. We are in some way a part of all this. So this perspective, rather than showing me that I am small and insignificant, instead tells me that I am *part* of something almost incomprehensibly grand. This sentiment is nicely expressed by Edwin Dobb:

> *Precisely when we grasp the vastness of the universe we also grasp an equally vast interior, the enormous geography of the soul, so to speak. Words may fail afterward, forcing us to rely on hackneyed descriptions that emphasize our insignificance, but what we actually sense, if only for an instant, is largeness of spirit.*[3]

With that narrative as a backdrop, it's useful to break things down a little more and just list some of the key topics and insights that might have an important bearing on our sense of place. These are some of the starting points, the building blocks of information about our context that we currently know. They are my tentative answers to the question, "For someone looking for personal meaning (where do I fit into the overall scheme of the universe, what makes my life significant, etc.), what would I list as the important insights from science so far, which this person needs to know as part of an effective search?" They are biased toward physics and astronomy due to my background and the focus on giving a broad sketch and illustrating the approach. Treat this list as a beginning, and an illustration of the types of insights we're looking for. Try to look at other topics, from other areas of science, in a similar way. In all cases our guiding principle is to seek the core insights that seem likely to significantly change our perception of our place in the world as we learn more about them.

The aim here is certainly not to teach you any of these topics; it's simply to list them with a brief explanation, pointing out some areas that might be important in your search. These subjects form the current structure of the courses and related information offered through the Science Integration Institute. Additional resources for exploring some of these topics can also be found in the bibliography.

As you build up your own favorite examples of insights from science that change your perception of the world, you'll gain more intuition to look at new discoveries and put them into your framework of meaning. A significant amount of background is necessary to make that possible, to even recognize enough of the framework to see how science can be used in this way. The examples that follow will help give you some ideas to start or continue building up this background.

Understanding the basic process by which the universe can be investigated through science – This is perhaps the most important topic of all. Without an understanding of the process by which scientific knowledge is gained and understood, it's very easy for any insights or facts you learn to become static and dogmatized. Whenever ideas and discoveries from science filter out into broader culture in a dogmatic way, without an understanding of their limitations and how they came to be known and understood, there is a tendency for them to be misinterpreted or overgeneralized in ways that discourage further exploration, refinement, and development of worldviews.

As we continue to learn more, and as you are bombarded by claims to knowledge from many sources, it's important to be able to interpret new claims with some sense of connection to how the system of learning works. Of course you can't personally check the evidence for every claim that is made, but if you know the general process, you'll be in a better position to filter, and to avoid viewing science as a set of abstract statements believed on faith in the authority of scientists as "experts."

How does one draw conclusions in science, to make the claims seen splashed across the headlines? How does one use evidence to build up a theory; how does the error-correcting process work to modify and improve a theory? Science is ultimately a process of choosing what works and discarding what doesn't work, kept honest by the awareness that someone else can come along and repeat your experiments and test your claims for themselves. It's a guided trial-and-error process not unlike the common sense process we all use to solve problems in daily life, but extended to new and often unfamiliar situations. An awareness of this will go a long way toward making the process and the results of science useful in your worldview-building effort.

> *It is neither possible nor necessary for the general public to have detailed scientific knowledge across a range of disciplines. Instead, what is important is scientific awareness—an understanding of what the scientific enterprise is about, what a scientist means by the word*

> *"theory," and what it means to establish a "scientific fact." For instance, many people say "evolution is just a theory," assuming this means its basic principles are still debatable. They do not realize that gravity is also "just a theory," and that, to a scientist, a theory is an explanation of what has been observed.*
>
> —Keith Devlin[4]

Overall structure of our cosmic context in space and time – It's useful to have a rough sense of time and distance scales for significant processes and structures in the universe. The universe is much bigger and much older than we can easily comprehend or relate to our ordinary experience, so it helps to have a few benchmarks to provide a sense of perspective.

For example, we live in the Milky Way Galaxy which is an island of a few hundred billion stars, and it takes light about 100,000 years to travel from one edge of our island to the opposite edge. There are many billions of galaxies similar to our own scattered throughout the known universe, typically a few million light years from each other. The space of which our universe is composed is expanding (*i.e.* galaxies are moving apart from one another). Our universe is roughly 14 billion years old—that many years have elapsed since the observable universe was compressed into a very hot, dense, tiny region. Our sun and Earth first condensed into their present compact forms roughly 4.6 billion years ago. So over nine billion years had already passed in the history of the universe before even the beginning of geologic history on Earth, the start of the processes that shaped the continents and mountains we see today.

Complexity and self-organizing properties of systems – Within the setting of this overall framework, systems form which organize themselves into various structures of greater and greater complexity. Knowledge of the ways in which they do this, the environments and basic rules by which such structures form and maintain themselves, is valuable for gaining perspective on how we came to be. As part of this perspective, it's also helpful to be aware of the origins of elements that serve as building blocks that now make up different materials and structures on Earth. Being able to trace the material of your daily life through the web of processes extending throughout the universe is a great way to enhance your sense of connection to the cosmos.

Key stages in the evolution of life – From our point of view, of course, life is a particularly interesting type of self-organizing, complex system. A basic understanding of the genetic basis of life and the connections and transition between living and nonliving matter therefore seems very important. The process of evolution by which life has changed and adapted itself to different environments also seems a vital clue to what's going on.

Also, just as in the cosmic and geological cases, some of the landmarks in the evolution of life will give us a sense of perspective, and show us how tiny a fraction of time is occupied by the period of human history we recognize as civilization. To make our relationship to cosmic history concrete, consider this: If your outstretched arms represent the entire 13.7 billion year history of the universe, with the beginning of the universe at the tip of your right middle fingernail and the present day at your left middle fingernail, then the length of time humans have existed occupies only a sliver of fingernail, less than a millimeter thick. Trim your fingernail slightly and you remove human existence from the timeline, but leave the vast majority of cosmic history untouched!

> *Four thinkers since Galileo, each informing his successor of what discoveries his own lifetime had seen achieved, might have passed the torch of science into our hands as we sit here in this room [in the late 19th century]. Indeed, for the matter of that, an audience much smaller than the present one, an audience of some five or six score people, if each person in it could speak for his own generation, would carry us away to the black unknown of the human species, to days without a document or monument to tell their tale.*
>
> —William James[5]

Mental processes and cognitive psychology – For our sense of what we're doing with our lives, consciousness is obviously a crucial element of the cosmos. While we may not be the central feature of the universe, it doesn't seem overly anthropocentric to note that it *is* very remarkable that such a thing as consciousness exists at all. Surely it is an important, significant feature of the universe in some way, which we may yet be a long way from understanding. Our beliefs affect us and the rest of the universe through their influence on the decisions we make. So a worldview that has any purpose to it must be anthropocentric in the sense that it influences *our* decisions. That's not necessarily saying that humans must be central to the universe; it's just recognizing that this is what we have control over. So anything that's going to affect what we have power to do at all must be centered on us and on our conscious awareness.

For now, some basic information of what we do know about consciousness seems valuable. What do we understand about how we learn, how we construct worldviews for ourselves? Obviously the mental maps through which we relate to the world are ultimately in our minds, so it is important to have some understanding of how these maps are constructed, how they might be distorted, how we can more effectively live according to the maps we consciously construct.

Basic laws of nature underlying everyday constraints and limitations we experience – At its most fundamental, our study of science could be seen as arising from the desire to understand the constraints and limitations we experience on what we can do and how we can do things. As discussed in Chapter 2, we can visit the moon, but only by working with specific rules and limitations that are imposed on us by an external world. We can't simply wish ourselves there. Many fundamental principles of science permeate our experience with the world, and this can be seen more clearly when we view science as a recognition of the fact that nature imposes limits and constraints on us. This perspective can help organize our thinking about what we are basically doing when we interact with the world. But we have to learn to consciously experience our surroundings at this level and to relate the science concepts and insights to what's going on in our lives.

When I buy an apple and eat it, how can I see this in terms of the process of energy coming from the sun, being stored in the molecules of the apple, and ultimately enabling me to be climb up a steep flight of stairs later that afternoon? I cannot go forever without food or water or without sleep, though I can choose to ignore hunger or tiredness for awhile.

The perspective this idea can offer on your daily life is well demonstrated by focusing on energy as a central concept. You can learn to trace through many of the concrete constraints you experience in life by looking at the scientific concept of energy as a central thread running through these constraints. An understanding of energy; what it is, how much of it we get from the sun, how much we need for different activities we want to carry out, how it connects different parts of the world, can help you connect together in your mind many pieces of the world that otherwise seem quite separate.

For example, we need to eat 2000 or so Calories each day to maintain our basic level of activity. This is because our activities (moving around, thinking, breathing, blood circulation, etc.) require power, just as the operation of a light bulb, toaster, or computer requires power. Our 2000 Calories per day translate into an average power of about 100 Watts, similar to the power consumption of a light bulb. The energy stored in the food we eat comes ultimately from the sunlight striking the surface of the Earth. So we can figure out how much of the Earth's surface is needed to collect enough energy to feed each person. Very roughly, this works out to about one square meter. As you consume your daily supply of food, think about the fact that you are making use of the sunlight falling on one square meter of Earth.

Technology – I don't mean this in the sense of keeping up on all the latest gadgets, which would be an overwhelming and probably a very distracting task. Rather, I'm talking about a basic understanding of the relationship between technology and our personal and social objectives. The development of technology is an integral part of the overall development of the universe, yet we tend to isolate and separate it. How can we integrate our personal and social sense of direction with the technological capability we gain in order to pursue these directions? Most of the time the conveniences of modern technology fade into the background of our daily lives. But by becoming more aware of them and focusing our attention on the basic principles of nature by which they operate, we can gain a deeper awareness of how our universe ticks. A radio you use to receive information transmitted through empty space can be a window to the deepest mysteries of existence, as surely as a night under a dark starry sky can be.

Core ideas related to the foundations of reality, space, and time – Several areas of physics that are often viewed as fairly inaccessible to the non-specialist have some very profound implications for our sense of what reality is, which we need to be aware of at a conceptual level. Some ideas from quantum physics reveal an underlying reality that is very different from many of our common sense notions, and concepts of Einstein's theory of relativity change our view of space and time. Clearly reality, space, and time are very important parts of our experience with the world, so some understanding of these ideas will be valuable in the development of our worldviews.

> *Quantum phenomena challenge our primitive understanding of reality; they force us to re-examine what the concept of existence means. These things are important, because our belief about what is must affect how we see our place within it, and our belief about what we are. In turn, what we believe ultimately affects what we actually are and, therefore, how we behave. Nobody should ignore physics.*
>
> —Euan Squires[6]

Reflection and Discussion

• Try to identify some of the ways in which you already live differently because you know specific things about the world that were learned through science. In other words, what are some concepts or pieces of knowledge for which you would say, "I make different choices, act differently because I know —"?

As we've discussed before, our choices are certainly based on many subtle influences, so in some cases it's hard to be specific. But it can be fun and enlightening to try to articulate clear instances where what we do seems pretty directly guided by beliefs we've appropriated about how the world works.

To get things started, one common example is the image of the Earth viewed from space. This tends to make us see the Earth as a whole, unified ecosystem, which influences many of our daily decisions about recycling, etc. See how many of these sorts of influences from scientific information you can come up with, in your own life.

• What information about the natural world do we use as the basis for our social goals?

> *[The objectives of the human race] have not been reconsidered in light of the science of the past few hundred years.*
>
> —Gerald Feinberg[7]

Chapter 6

Directions for Exploration

Summary: This is an overview of the flavor of exploration that investigates nature with an eye for uncovering insights about our role in the universe. It's also a sampling of some areas of research I think are likely to uncover insights that will be particularly valuable in our worldview-building efforts. So these are a few topics you might want to keep up on in the future. Most importantly, my discussion of these topics is meant to convey the spirit of this type of research, which could be applied to many areas of science as well.

We would shift our emphasis from trying to discern the structure of the universe to trying to reckon our place within the structure...

—Edwin Dobb[1]

Before I describe the sort of research I have in mind, let me try to convey the flavor of it with a short reflection.

Like many people, I love music. I like being able to experiment with the combinations of sounds and their influence on the way I feel. How do all these different sounds combine together to make me feel a certain way and to give me a connection to things that I didn't feel just a moment before, when everything was exactly the same except for a few vibrations now going through the air? If I turn on some music as I sit here typing, almost nothing has changed, really; nothing that a casual observer would see as different about the setup at all. I'm still sitting, in the same position, in front of the computer, with my desk arranged the same way. Yet in my mind, everything has changed. My mood is different, my thoughts are totally different, just because the vibrations in the air have changed slightly.

Of course it's not really the vibrations directly that cause the change in my mood; it's the information they carry, the experiences they represent to me. But that's what is so wondrous about nature, and it points to the spirit of what we're really trying to uncover in our investigation of nature: What is the relationship between the physical constraints, the physical representation that is necessary to make anything happen, and the information, the meaning, that makes these physical arrangements matter to us? We don't really care directly about the physical arrangements; we care about how they feel, what they mean. How are all these things intertwined to make life what it is? We want to look at the universe as a system that somehow allows the desire for meaning (as expressed through us) to arise. Our aim is to investigate the universe from this perspective.

If you conducted an informal survey of scientists about why they do their work, you'd get many different answers. All scientific research, like any other human endeavor, is carried out with some objective in mind, but this objective certainly varies from project to project and from scientist to scientist. The objective might be the production of technological advances, a vaguely defined pursuit of pure knowledge, the quest for fame and recognition, or one of many other goals that scientific research helps serve. In any case, the types of questions asked, and the culture of the field of study, are strongly influenced by the objective.

> *But nature gives most of her evidence in answer to the questions we ask her. Here, as in the courts, the character of the evidence depends on the shape of the examination, and a good cross-examiner can do*

> *wonders. He will not indeed elicit falsehoods from an honest witness. But, in relation to the total truth in the witness's mind, the structure of the examination is like a stencil. It determines how much of that total truth will appear and what pattern it will suggest.*
>
> —C.S. Lewis[2]

It seems odd, and reflects the split that has occurred between science and the search for meaning, that the direct support of efforts to build meaningful worldviews for ourselves is not a commonly cited objective of scientific research. We have research efforts focused on a tremendous variety of important social and personal goals: cures for medical conditions, cleaner energy sources, travel into space, faster computers or more impressive graphics for movie production, and so on. But rarely do we find mention of research guided by the conscious objective of providing information people can use in seeking meaning in their lives, and where the presentation of research results is focused on interpreting what is learned to help build a picture that influences our perception and our sense of place in things. As we've discussed in Chapter 1, these pictures or worldviews play a dominant role in our lives, so surely this impact of science is important, and there is a need for some work in science that is directed by this focus. The research described here is unified by this common objective. The questions we ask of nature are guided by the theme of trying to uncover a clearer perspective of our role, as conscious creatures, in the overall scheme.

This type of research is not a well-defined category, and I make no attempt to be precise or limiting in defining at this early stage what topics best fall within this category. In some ways it is for now only a question: How can we investigate nature in a way that starts with the search for meaning, and seeks to use the methods of science to uncover key insights about the role we play in the universe? The ideas below are just examples that might help illustrate the spirit of such an approach to research. I hope this will also serve as a guide to a perspective you can take on new developments in science, and to some areas you might wish to follow more closely or learn about in your personal pursuit of understanding and perspective.

In understanding the scope and aim of this broad category of research, it's helpful to return to an analogy from Chapter 2, in which we imagined finding ourselves on a stage during a performance, looking around to try and figure out what was going on and what part we played. The aspect of scientific research we're discussing here amounts to looking at *nature* in this way. A wide variety of seemingly unrelated areas of science may contribute to this effort. The unifying theme that ties them together is the investigation of nature with the specific aim of generating insights that help us see how we fit in, what role we play. In principle, any new information from any area of research in science might be valuable in this effort. What really matters is the perspective from which the research is approached and the kinds of insights we try to extract as we gather new information. So with the right approach, nearly any area of scientific research could fall within the realm of what we're looking for. I encourage you to consider other areas of science in a similar light, asking what areas of research seem most valuable to you in the effort to establish a clearer connection and a clearer place for us in the maps provided by science.

Suggested reading to get started on investigating some of these topics—many of which are extensions of topics discussed in Chapter 5—may be found in the bibliography at the end of the book.

Scientific cosmology – Describing and investigating the physical arrangement and history of the universe is important because it clarifies the overall framework in which everything else happens. It is the investigation of our context in the largest sense. Cosmology has benefited in recent years from rapid advances in the ability to collect important data about the universe, so it seems likely that we will see many new insights from this field that could profoundly impact our personal worldviews.

The anthropic principle in cosmology – For the purpose of developing our worldviews, we're particularly interested in cosmology as a way to understand the processes that had to happen to make consciousness possible. What properties did the universe need to have in order to make it possible for us to be here wondering about it at all? And given all the possible ways the universe could have been, why does it have the right properties to allow us to emerge? This observation is related to the "anthropic principle," which explores the extent to which the fact of our existence helps explain why the laws of nature are the way they are.

Discussion of the anthropic principle opens the door to questionable science in some cases, so we need to tread carefully here. Nevertheless, questions about what type of universe is capable of supporting life, and how likely or unlikely it was for our universe to generate the proper conditions, are certainly important. For the purpose of thinking about how science affects your worldview, they are important primarily for shifting the focus of questioning onto our place in the universe, rather than just on the universe in general.

The basis for the arrow of time – Our everyday experience is dominated by a perception that the future is fundamentally different from the past. We remember the past but not the future, and we have the ability to change the future but not the past. The sense of urgency in our lives and our ability to feel pride and regret about what we have done are inextricably tied up with the fact that we live in a universe with this fundamental property of irreversibility. Yet it is unclear just how this dominant aspect of our experience connects with the fundamental laws of physics, which for the most part do not make a distinction between past and future. Work in this area is important in order to provide a bridge connecting the world of ordinary experience to the objective and abstract world described by physics.

Foundations of quantum physics – Quantum physics reveals deep puzzles about the nature of reality. Many of these puzzles are far from settled, and a familiarity with some of what's going on in this field will certainly keep us from becoming too complacent about what we think we know of the reality we all share. One of the most puzzling aspects of quantum mechanics is the role of measurement in the theory. When a measurement is made, a state that previously was described only by probabilities for various outcomes now must pick one specific outcome. Although it's well understood how to predict the results of a measurement based on the rules of quantum physics, the question of what makes one process a "measurement" of an observable, while another process does not constitute a measurement, is not so well understood. The rules of quantum mechanics seem to require a clear dividing line between systems that can be observed and analyzed in terms of the quantum formalism, and systems which act as classical "measuring devices" and cause the system to choose an outcome. Yet, there is no clear place to draw the distinction; a measuring device seems to be just a collection of many of the same kinds of simple systems to which quantum mechanics can be applied. There are many possible solutions that have been proposed to resolve this difficulty, but it remains a subject of debate. In any case, an understanding of quantum foiundations might contribute to your understanding of other concepts important in the search for human meaning, such as free will and the arrow of time, so some awareness of new developments in this area will surely stir up your thinking and broaden your perspective.

> *In the last analysis, can a satisfactory description of the physical world fail to take explicit account of the fact that it is itself formulated by and for human beings?*
>
> —A.J. Leggett[3]

Links between information theory and other sciences – This is the topic I was pointing toward in the reflection which opened this chapter. Our worldviews and sense of meaning of course operate in terms of information, because they consist of ideas. But the information used to express ideas is stored and changed through its representation in some physical form: scribbles on a sheet of paper, electrons in a computer, molecular structures in our genetic code, etc. There is much interesting and ongoing work in molecular biology, thermodynamics/statistical mechanics, quantum theory, and other areas involving the interface between information and its physical representation. I think this a particularly promising area for further exploration.

Cognitive science/psychology – As we discussed in Chapter 5, improvements in understanding how we appropriate information into our worldviews are obviously very important. The process of incorporating our knowledge of the universe into the perspective from which we live each day involves a very different style of learning than just memorizing abstract facts. There is much ongoing work in this area that can provide new ideas to try in your own efforts.

Complexity and self-organization – The relevance of this topic to our efforts was discussed in Chapter 5. This is a relatively new science, so many of the core insights we'll want to be aware of are still very much under development.

Research directly impacting our sense of what it means to be human – Rapid advances in areas such as biotechnology (e.g. genetic engineering and cloning) and computing power (artificial intelligence, etc.) are already starting to turn previously philosophical discussions about what it means to be human into pressing, pragmatic issues.

Historical origins of the separation between science and meaning – A better understanding of how and why the rift emerged in the first place will provide insight into how to repair it. At the same time, it will help make sure we don't discard important aspects of the scientific method when we apply it to research questions whose answers may provide insight about human meaning. Of course, much work has been done on the history of science and the attitudes of society about science. Work in this area continues, and new developments will likely help shed light on our own individual thinking about these questions, and help us focus on linking this work to the practical questions of how to change attitudes and reconnect science and meaning.

Understanding why people reject scientific ideas – In allowing insights from science to be more readily assimilated into worldviews, it is important to learn exactly what we may sometimes be looking for that we think a scientific worldview cannot provide, and which we may turn to "pseudosciences" or other sources to find. In many cases, I think, we are mistaken in the belief that we must reject scientific information about the universe in order to find what we're looking for. Thus there is value in trying to more clearly identify some of the key features of what we are looking for, understand why we think these features are incompatible with a scientific view of the universe, and evaluate whether they may in fact be compatible (though perhaps in a modified form) with what science has revealed about nature.

In ending this chapter, I'd like to repeat that these were just a few examples to illustrate an approach to investigating the universe which is guided by the search for insights and perspective into our place and role. Please use the examples as a guide for identifying areas that matter most to you, and incorporating new insights from these areas of science into your search.

Reflection and Discussion

• What areas of scientific research seem most interesting and exciting to you? What do you follow and look for in the newspaper, Internet, or TV? Try to zero in on these topics and consider exactly what makes them so interesting to you. Which topics seem to give you the feeling of being connected to the mystery of things?

• Look back at your answer to the last Reflection and Discussion question in Chapter 3 (What new facts could be discovered that would strike you as evidence that this is a meaningful universe?). How might different areas of research connect to the answer you gave, to help provide you with the information you seek? Which kinds of research seem most important to you, on that basis?

Chapter 7

Toward a New Relationship with Science

Summary: This final chapter summarizes and reemphasizes the vision of what things might look like if we achieved the relationship with science I've been building and advocating throughout the book. It also recaps some of the challenges we face in trying to live in a way that maintains conscious connections to the broader universe, and invites you to accept the challenge and embark on the amazing adventure of making your daily life more meaningful and connected to the cosmic processes you are a part of. Ultimately, the value of learning and understanding the universe is in how we express that understanding through the way we live each moment of our lives.

We study the story [of the universe] primarily in order to live the story.

— Brian Swimme[1]

Let's review where we stand now. We recognize that as individuals, we seek to guide our lives by some context that gives meaning and significance to our decisions and actions. Perhaps the essential problem of life boils down to the need to feel that it matters, at this instant, what we decide to do next. We need to believe we are not helpless. We'd like to believe that there is some meaning and some kind of real "fight," in connection to which it matters that we muster our strength and make choices that seem right in some well-grounded context. So we struggle to make that context intellectually honest—to reflect accurately the external reality we all experience and share. Uncovering insights about this external reality falls within the domain of science. We need to formulate for ourselves a relationship with science that is not alienating, which puts us on its map and invites us to consider how its insights affect our worldview.

A scientific description of the context we live in does us little good unless it makes explicit connection to the immediate, pressing decisions we face in everyday life. Many of these decisions seem small and prosaic compared to the grandeur of a vast universe encompassing billions of light years in space and billions of years in time. Knowledge of the context which has made our existence possible gets its power only when it can be translated into a form that impacts the countless tiny decisions that together make up our lives, our societies, and all that we do as humans.

> *Small as we are, minute as is the point by which the cosmos impinges upon each one of us, each one desires to feel that his reaction at that point is congrous with the demands of the vast whole.*
>
> —William James[2]

For many people, science has had a negative impact on their search for meaning and purpose in their lives. As discussed in Chapter 4, scientific investigation has revealed information that has been destructive of many sources of meaning, revealing some of the beliefs associated with these worldviews to be inconsistent with reality. But I'd like to convey a much more optimistic, hopeful message. Is it really so surprising that learning more about the world should reveal flaws in our conception of it? Why take this in such a negative way? If we truly believe in meaning, we should rejoice in finding helpful tools in our search, a long search, to bring us closer to what this meaning might actually be.

In the previous chapters I have tried to build up the argument that insights from science are important tools for helping to construct the worldviews that ultimately guide our lives. Our worldviews dominate our choices and our lives, even when we've nearly forgotten these worldviews exist. A greater awareness of the insight-building capacity of science will help us to make more complete use of its tools. This awareness consists of four key elements:

1) Our lives boil down to the choices we make.

2) We construct maps for ourselves that guide these choices.

3) Science has much to say about the maps we want to make (if we think it matters to make good maps).

4) We must make a conscious effort to extract information from science in a way that will help do this—the insights will not fall naturally out of most types of research, and this effort is very different from just focusing on the technological implications of science.

One component of this effort will need to occur within parts of the scientific community. We'll need to continue working to present and practice science in a way that invites and encourages people to naturally assimilate its key insights as part of their individual worldview-building process. Some scientists are working to do this, and change occurs gradually. It is perhaps not so surprising that we haven't yet done this very well. Even some of the most basic and fundamental insights from science are less than a few hundred years old, spanning very few generations of people. There hasn't yet been time to really assimilate many of the ideas into our perspectives. But the good news is, you don't have to wait for any of this change, in order to get started on your own. Science truly is available to everyone, everywhere. All you have to do is look around, and ask questions. You can start making real, concrete progress in connecting your life to the broader context, in making your life more meaningful (in whatever way you define meaning for yourself), beginning right now. You can start being bolder in thinking about the implications of the science as you learn it, using what you learn to direct your questioning process and ask seriously, "What could be a meaningful place for me, given what I know? How can I allow the science to touch the problems of my life, help me build a map that puts myself on it in a way that is satisfying to my inner need for meaning, but is also consistent with what we learn through science about the universe?"

So in this final chapter I'd like to bring us back to action and a direct link to what we do with each minute of our lives. All the theoretical understanding in the world is meaningless if it doesn't become a part of who we are, how we live each instant. These everyday instants, after all, are the real stuff of life. They define us and characterize who we are and how we impact our society and our world. And they are all that we can ever directly control. The greatest, most heroic accomplishment we could ever imagine for ourselves still must be achieved by piecing together actions selected from the choices available to us each day. If anything we accomplish really matters, then these everyday choices must matter. Here I'd like to suggest an operational framework for moving forward. We need a concrete way to make progress in small, manageable steps, starting from where each of us is right now, toward objectives that are part of our immediate lives, but are grounded in a much bigger context.

Here's a broad outline of this framework—the details are of course unique to each individual. I sense that our own feelings of uncertainty about what we're doing here reflect something very fundamental about our universe: a universe that explores, that is working toward something, but is "uncertain" and is trying things out as it goes along. Now, very recently, beginning just a million years ago or so, the universe has perhaps had a breakthrough on this tiny little planet. It has created a fundamentally new aspect of itself that has become aware of this uncertainty, and is able to create ideas of how things might be, independent of the truth or falsity of the idea in reality. We can dream; we can hypothesize; we can formulate models and stories of what we're doing here. And we can try them out, see how they work, and modify them as we learn more. We do this with our form of government, with our school systems, with our choice of college major or choice of career, with our choice of friends, with everything about our personal and social lives. But like the universe we are a part of, we seem to have no direct line to truth. All we can do is keep trying things out, and trust the feelings that come to us every once in a while, that tell us there is *something* behind all this, perhaps only dimly perceived, but still something that matters and is worth striving for.

So just get started. Look at your own situation. Look at what's going on in your life. Start thinking about how to put yourself in a position to believe, concretely and honestly, that there is something fundamentally worth doing, that is within your power to do. How can you begin to organize your activities around putting yourself into a state of mind in which you fully believe you are doing what matters, and are working to gain the capability to do things that matter?

As we go through a day, we're continually evaluating our feelings and inclinations to do things, labeling some as "good and valuable," to be followed, and others as "bad" or irrelevant or somehow not valuable to act on. Pick out a few of these things, a few of these feelings or choices, and start tracing them back to their origins. What objectives do you have in mind, that lead you to think of them as good or bad? Your personal worldview from Chapter 1 might be helpful here as a starting point. What assumptions are you making about how the world works, in deciding which things you will or will not do? What assumptions are you not so sure about, that you'd like to investigate further or learn more about? What information might you need to build a solid worldview around these things, a background context in which they get their meaning? Some examples to get you started are the job you're working in, where you live, what products you buy, what food you eat, whether and how often you exercise, the people you like, and what hobbies you enjoy.

After you've identified some of the important things going on in your life, start thinking concretely about how they fit into what else is going on in the world. Just let yourself try out scenarios as real, direct attempts at considering what it means that the universe has the properties we observe it to have. You can start very small with things you directly observe. Why do things consistently fall when you drop them? Why are there so many different life forms around? But you can also take advantage of what many generations of explorers have uncovered, read about things they have learned, and use those insights. In any case you can always start from where you are right now, and ask, "What does it mean, to me and to my sense of place in things, that some property of the universe is like that?" Everything has a context. You are in charge of constructing your own understanding of that context, and you can begin *now* to make progress with each observation, each thought, each connection you make. You don't need to get everything "right" from the start. The important thing is to have a starting point, a way to move forward. The scenarios you generate are working worldviews that place some of these things into a context and help you find out what you need to know next, while giving you a guide to start with for now. And by building and adapting your worldview out of questions from daily life you ensure that the worldview you develop will be of immediate, practical use to you.

As you collect together some of these ideas about what seems important to you, combined with information from science about how the world seems to work, you can start looking for patterns to give you some insight into where you think you're trying to head with your life, and how this might fit into some sort of meaning that could be what's going on in the universe. Then you can generalize to give yourself tentative guiding principles for how you want to live, to guide your immediate decisions. For example, a very simple description of a worldview for giving meaning to your life might consist of these three elements: 1) Here are some rules to guide your behavior, stated in absolute and unchanging form. 2) Here are the feelings and impulses you will experience, some of which will push you toward, and some away from, the rules. They are largely irrelevant to the rules (or may even work in direct conflict to the rules). They are just impulses that will make you feel inclined to do certain things. 3) Your goal is to follow the rules, regardless of the impulses.

An alternative scenario, again expressed in simple terms, might suggest that the various impulses we feel are a collection that has been built up over a long process of evolution, with new impulses put down on top of old ones as we developed and circumstances changed. Perhaps some of the old impulses, still buried within our makeup, were valuable in the past, but may be outmoded now if they developed in response to circumstances from long ago. Thus we feel the conflict of suppressing some impulses and encouraging others, based on knowledge that can adapt much more quickly than our instincts.

It's worth asking whether you believe either of these basic models, or some more complex version of them. It's also worth noting that your understanding of science and the perspective it gives on how we got to our current state certainly has a significant impact on how you feel about these scenarios.

Once you get the hang of looking at your daily life in this way, you'll be able to stop and ask yourself, "What am I doing (or thinking) right now?" and answer in a way that goes far beyond the activity itself, putting the activity into a context of meaning that you are gradually building up for yourself. Imagine living most of your life with a conscious awareness of your context, of the network of processes that sustains you and puts you where you are. This can bring a greater awareness of how your own choices and actions right now will affect this network, hopefully bringing greater meaning into the world. We tend to lose our sense of wonder and our sense of motivation when we are too narrowly focused on our immediate surroundings—when we fail to see the larger web that our action fits into. Imagine having a clear worldview for guiding your choices into harmony with the overall scheme of things, for filtering the activities you engage in on the basis of whether you can see them contributing to something bigger.

From this perspective, life events take on a new meaning. I'm not just filling out income tax forms; I'm participating in a flow of information that allows resources to be organized in certain ways to further certain social goals that are also part of the cosmic processes. I'm not just falling in love; I'm expressing powerful feelings that are a product of the stars, the sun, the Earth's ecosystem, the development of consciousness, etc., and which express something meaningful about all these processes. I'm not just planting seeds in the ground to get that over with and move on to the next task. I'm working within the constraints of the universe to tap into a process that enables food to grow by harnessing the energy of the sun, so I can have energy for carrying out other activities that are a part of my worldview. We often carry out actions we perceive as pointless or irrelevant, because we think we somehow "should" do them. Having a clear worldview helps us remove things from the "should" category, and put them solidly into either the "valuable" or the "irrelevant" category.

The challenge is to think about your actions and decisions in the context of these scenarios, and to engage in the continual process of modifying and adapting your scenarios as you learn new things. Imagine how wonderful, powerful, and meaningful it would be to be able to ground every decision and every action in a scenario (consistent with what science has revealed) describing what the universe is doing and how your individual action fits in.

The development of a worldview and the use of it to make major decisions can still seem a little abstract. People don't generally ask themselves, "Why am I an electrical engineer?" or, "Why do I recycle?" every day. So here are a few points about how this change in thinking can directly impact the way you live:

1. When worries arise in your life, your worldview can give you a concrete framework for asking yourself what the feelings and thoughts are based on, so you can make some sense out of the feelings and gain enough perspective to not let them control you. You have a framework for asking, "Are my concerns really important? Do I need to be worried about this, in the bigger context of things?" You can place yourself within your

overall context (as stated or modified from the end of Chapter 1) and ask what your worldview says about this situation.

2. Understanding your own worldview and the assumptions it is based on tends to make you more tolerant of other people. Ultimately disagreements come down to different core assumptions that go into a worldview. Bringing these points out into the open allows you to have an intelligent and fruitful discussion about the validity of the core assumptions, rather than nitpicking about conflicting details that are really just a necessary, logical consequence of the core assumptions. No real progress is ever going to be made in the argument until you are able to move to a level at which you address the core assumptions and investigate their validity. Otherwise, you'll just each be stating your opinions, from within different, incompatible, worldviews.

3. Having a well-articulated worldview makes you more confident or at least decisive about your actions. Even if you end up being "wrong," you can explain and defend why you did what you did. You'll know it made sense according to the best framework you had to go on at the time, and that's really the best you can be expected to do. Articulating your worldview forces you to be concrete and deliberate about the foundations for your preferences. Often we feel conflicts because we don't have a clear framework for deciding which thoughts or actions matter and which do not, so we act half-heartedly. We can do something, with a part of us knowing all the while that we don't really think this is important in the overall scheme of things. We

watch television, knowing we might be better off doing something else, but without having a clear enough idea about how this something else would matter, to motivate us to do it (or on the other hand, without having a clear enough idea of how watching television might matter in the scheme of things, to stop feeling guilty and enjoy the full benefit of watching the show). Spelling out your worldview and relating it to what seems to be really going on in the universe allows you to ask, "Do I really think it matters that I do —? Does it serve some overall objective for the universe, as I currently understand that objective?" Imagine the power and sense of motivation that can come from living your life in this way. For each action, you will either be fully aware of how it matters because it fits into some part of your worldview, or you'll be aware that it does *not* fit, so you'll have the confidence to decide not to do it. You'll be much less likely to do something just because of convention or because of pressure exerted by others through guilt or other means. You can simply ask what your worldview says about the choice. If it fits in, you'll be able to do it, wholeheartedly. If not, you'll be able to confidently turn it down. Of course, it's not quite that simple, and we're constantly modifying and updating our worldviews even when we are fully conscious of them. But at least it gives you a solid framework to operate from.

4. It's liberating to realize that you actually can explore and experiment to build up your worldview. You realize we don't know the answers, but you can try out possibilities and make real progress even without knowing absolute answers. Life is a wonderful adventure!

5. You're a much better problem solver when you have this perspective. You more easily "get to the bottom of things" (i.e. to the fundamental level). You're calmer because you have a sense of perspective about the problems you face in life, and you aren't constrained by conventions. You see the assumptions on which conventions are based and can investigate them directly, to see if circumstances have changed so that actions which may go against these conventions are actually better serving the overall objectives you perceive.

These pages have probably given you many more questions than answers. I think this is as it should be, since we are near the very beginning of what will likely be a long search to understand our role in the universal context, and the answers may well be different for each individual. We're still seeking directions in which to set out to even begin to look for some of the answers. So I leave you to consider again the questions that I think express the essence of the suggestions I've offered for finding such a direction:

Do we believe that behind all the concerns, struggles, and details of daily life, there is something that truly matters in some way? Does it make a meaningful difference what we decide and how we act? Surely we don't yet know with any certainty *how* it matters, but do we *think* that in some way it does matter? You of course are free to answer that it does not, in which case the search (and this book) are unlikely to interest you. It's not necessary to believe that anything we decide or do actually matters at all. But most of us certainly *behave* as if it matters a great deal. Any argument we make or strong belief we hold about anything implicitly assumes some background against which it does make a real difference what we do. So for those of us who are persuaded by the gut feeling that it is significant, somehow, what we do and think, we are left with another question: How can we organize a productive search that will make progress in understanding what matters, and how it matters? Don't we need the insights and methods provided by science, in order to enable us to gain enough context about what's going on in the universe to be able to search effectively? My hope is that through a new relationship, a new partnership between science and each individual looking to live a significant, meaningful life, new opportunities for progress in this search will be opened. We need the drive to find out what matters, the need for meaning, which comes from within each of our hearts. But we also need the power of science to filter claims about the world, to distinguish ideas that work from those that do not work.

In any case, for each of us with open eyes and an open mind, the search seems sure to be an exciting journey, revealing a world that is far from ordinary. I wish you a journey filled with wonder, surprise, and meaning.

Reflection and Discussion

• What can you do right now to start living more actively and consciously "in the universe" as you perceive it and express it through your worldview? Set alarms or arrange other reminders a few times a day to stop and consciously ask yourself, "What am I doing right now—what am I *really* doing?" Connect your activity as widely as possible to the rest of the universe. What system are you a part of, what objective are you working toward, and how might that objective fit into a bigger scheme that it contributes to? Can you tell a story about what you're doing that connects it to the Earth and the sun and the stars and the rest of the stage as you perceive it? If you can learn to do this, you'll be in a much better position to make choices and live deliberately, knowing you're living in a way you could defend to the best of your knowledge, as the best you could possibly be expected to do.

• Make a list of what seems most important or valuable to you in your life: morals, accomplishments, friends, family, experiences you want to have, causes you want to support, social changes you want to encourage, etc. Then look for common themes in what you value, articulate these themes, and write them down. Now try to trace the foundations of these things in how the world works. What do your values say about what you think might be going on overall in the universe? What might a concrete scenario look like, that incorporates your values? See what information you need to build a solid worldview around them, a background context in which they get their meaning. See what new information you'd like to have about the world in order to clarify or expand or modify this context. We're always feeling conflicts over the choices we must make, knowing there is not time to do everything that crosses our minds as "yeah, I'd like that or would like to do that." So it's valuable (though difficult) to concretely try and establish scenarios for a real foundation of what's going on in the universe that could make your preferences ultimately valuable. For each of your preferences that does have a solid grounding, there *must* be some such scenario—we just don't usually force ourselves to articulate it. Try to make the scenario real and conscious and concrete for yourself. Often the things that are really important to you will emerge more clearly in the process of doing this, and those which are not so important can be more easily dropped. For example, if you think it is important not to hurt anyone, what could be behind things that would make this a natural part of what the universe is doing? How does this mesh with the properties we have learned about the universe through science?

If you're feeling really ambitious, also try to find a place in your worldview for the things you decide *don't* really matter. What story can you tell about why they exist, why you feel tempted to do them, if they are not really important things to do?

• From all these lists you've been making, try to extract a guiding principle; something to direct your choices in a simple, easy-to-remember way. Then you can continue to work more abstractly on grounding this in terms of your worldview, but in the meantime you'll have a simple way to guide your daily choices, knowing that deeper thinking is behind it, but without having to recreate that thinking at every decision point (good luck trying to think through your whole worldview for every decision!). As a simple example, asking, "How will this action help the universe more fully experience the feeling of magic and wonder?" could be a guiding principle.

• Suppose you somehow knew with certainty that there was absolutely no meaning in the external universe—nothing outside of the meaning you create for yourself. What would you do in that case; how would you choose to live? You would still have the feeling that some things matter more than others; you would still make choices; you would still be faced with daily life just as you are now. What might you decide on as the things that matter? By imagining how you might create meaning from your current situation, right now, in the absence of any external standard, you free yourself to really think about what is necessary for the kind of meaning you value, and to then look for it in the real universe.

• What would you most like to change about your current situation? Why? This is a key piece of a worldview—we act primarily to change conditions we find ourselves in, so focusing on why we want change and how we achieve it provides a good starting point for increasing awareness of how we believe we relate to the rest of the universe.

Notes

Introduction

1. Quoted in Bryan Appleyard, *Understanding the Present: Science and the Soul of Modern Man* (New York: Doubleday, 1992), p. 15.

Chapter 1 – Worldviews: Placing Your Actions in a Bigger Context

1. Peter M. Senge, *The Fifth Discipline: The Art and Practice of the Learning Organization* (New York: Doubleday, 1994), p. 175.

2. Jeffrey Sachs, *The End of Poverty* (New York: Penguin Press, 2005), p. 1.

3. Louise B. Young, *The Unfinished Universe* (New York: Simon and Schuster, 1986), p. 185.

4. Viktor E. Frankl, *Man's Search for Meaning* (New York: Washington Square Press, 1985), p. 136.

Chapter 2 – What Does Science Have to Do with Your Worldview?

1. Connie Barlow, "The Way of Science: Exploring the Foundations of Humanism." *The Humanist* 58 (March–April 1998), p. 17.

2. Viktor E. Frankl, *Man's Search for Meaning* (New York: Washington Square Press, 1985), p. 132.

Chapter 3 – Applying this Perspective in Your Own Life

1. Colin Wilson, *The Philosopher's Stone* (New York: Crown,

1971), p. 87.

2. Walt Whitman, *The Portable Walt Whitman* (New York: Viking, 1973), p. 214.

Chapter 4 – Why Isn't Science Already More Integrated into Our Lives?

1. Edgar A. Poe, *Works of Edgar Allan Poe: Complete and Unabridged* (New York: Outlet, 1985), p. 721.

2. Quoted in Louise B. Young, *The Unfinished Universe* (New York: Simon and Schuster, 1986), p. 183.

3. William James, *The Will to Believe and Other Essays in Popular Philosophy* (New York: Dover, 1956), p. 61.

4. Bryan Appleyard, *Understanding the Present: Science and the Soul of Modern Man* (New York: Doubleday, 1992), p. 14.

5. Viktor E. Frankl, *Man's Search for Meaning* (New York: Washington Square Press, 1985), p. 141.

6. D. A. Rees, "Commentary: Time for Scientists to Pay Their Dues." *Nature* 363 (20 May, 1993), p. 203.

Chapter 5 – Surveying Our Context

1. Gerald Feinberg, *The Prometheus Project: Mankind's Search for Long-Range Goals* (Garden City, New York: Doubleday, 1969), p. 22.

2. Quoted in Art Hobson, *Physics: Concepts and Connections* (Englewood Cliffs, New Jersey: Prentice Hall, 1995), p. 29.

3. Edwin Dobb, "Without Earth There is No Heaven: The Cosmos Is Not a Physicist's Equation." *Harper's* 289 (Febru-

ary 1995), p. 40.

4. Keith Devlin, "Guest Comment: Rather than Scientific Literacy, Colleges Should Teach Scientific Awareness." *American Journal of Physics* 66 (July 1998), p. 559.

5. William James, *The Will to Believe and Other Essays in Popular Philosophy* (New York: Dover, 1956), p. 53.

6. Euan Squires, *The Mystery of the Quantum World* (Bristol: Institute of Physics, 1994), p. xiii.

7. Gerald Feinberg, *The Prometheus Project: Mankind's Search for Long-Range Goals* (Garden City, New York: Doubleday, 1969), p. 17.

Chapter 6 – Directions for Exploration

1. Edwin Dobb, "Without Earth There is No Heaven: The Cosmos Is Not a Physicist's Equation." *Harper's* 289 (February 1995), p. 40.

2. C. S. Lewis, *The Discarded Image: An Introduction to Medieval and Renaissance Literature* (Cambridge: Cambridge University Press, 1964), p. 223.

3. A. J. Leggett, *The Problems of Physics* (Oxford: Oxford University Press, 1987), p. 145.

Chapter 7 – Toward a New Relationship with Science

1. Brian Swimme, *The Hidden Heart of the Cosmos: Humanity and the New Story* (Maryknoll, New York: Orbis, 1996), p. 25.

2. William James, *The Sentiment of Rationality* (page 329 in *The Writings of William James: A Comprehensive Edition*, by John

J. McDermott. Chicago, IL: University of Chicago Press, 1977).

Bibliography

This is by no means an exhaustive list, but it may help get you started on your investigation of topics that seem most important to your own search. Related resources may also be found at the Science Integration Institute web site, www.scienceintegration.org.

Appleyard, Bryan. *Understanding the Present: Science and the Soul of Modern Man.* New York: Doubleday, 1992.

Baggott, Jim. *The Meaning of Quantum Theory: A Guide for Students of Chemistry and Physics.* Oxford: Oxford University Press, 1992.

Balashov, Yuri V. "Resource Letter AP-1: The Anthropic Principle." *American Journal of Physics* 59 (December 1991): 1069-1076.

Barlow, Connie. "The Way of Science: Exploring the Foundations of Humanism." *The Humanist* 58 (March–April 1998).

Barrow, John and Frank Tipler. *The Anthropic Cosmological Principle.* Oxford: Oxford University Press, 1986.

Brown, J.R., Paul C. Davies and J. Brown (eds.). *The Ghost in the Atom: A Discussion of the Mysteries of Quantum Physics.* Cambridge: Cambridge University Press, 1986.

Burke, James. *Connections.* Boston: Little, Brown, 1978.

Burke, James. *The Knowledge Web: From Electronic Agents to Stonehenge and Back.* London: Simon & Schuster, 2001.

Capra, Fritjof. *The Turning Point: Science, Society, and the Rising Culture.* New York: Simon and Schuster, 1982.

Carey, Stephen S. *A Beginner's Guide to the Scientific Method.* Belmont, California: Wadsworth, 1998.

Chaisson, Eric J. *Cosmic Evolution: The Rise of Complexity in Nature.* Cambridge, Massachusetts: Harvard University Press, 2001.

Chalmers, David. *The Conscious Mind: In Search of a Fundamental Theory.* New York: Oxford University Press, 1996.

Christian, David. *Maps of Time: An Introduction to Big History.* University of California Press, 2005.

Coveney, Peter and Roger Highfield. *The Arrow of Time: A Voyage Through Science to Solve Time's Greatest Mystery.* New York: Ballantine, 1992.

Dalai Llama, *The Universe in a Single Atom: The Convergence of Science and Spirituality.* New York: Broadway Books, 2005.

Davies, Paul C. and John Gribbin. *The Matter Myth: Dramatic Discoveries that Challenge Our Understanding of Physical Reality.* New York: Touchstone Books, 1992.

Dennett, Daniel C. *Darwin's Dangerous Idea: Evolution and the Meanings of Life.* New York: Simon and Schuster, 1995.

Dobb, Edwin. "Without Earth There is No Heaven: The Cosmos Is Not a Physicist's Equation." *Harper's* 289 (February 1995): 33-41.

Donovan, M. Suzanne, John D. Bransford and James Pellegrino (eds.). *How People Learn: Bridging Research and Practice.* Washington, D.C.: National Academy Press, 2000.

Duncan, Todd and Craig Tyler. *Your Cosmic Context: An Introduction to Modern Cosmology.* San Francisco: Pearson Addison Wesley, 2009.

Duncan, Todd. "Guest Comment: Making Physical Science Courses Valuable to Nonscientists." *American Journal of Physics* 65 (May 1997): 365-366.

Einstein, Albert. *Relativity: The Special and the General Theory.* New York: Crown, 1995.

Feinberg, Gerald. *The Prometheus Project: Mankind's Search for Long-Range Goals.* Garden City, New York: Doubleday, 1969.

Feynman, Richard P. *QED: The Strange Theory of Light and Matter.* Princeton, New Jersey: Princeton University Press, 1988.

Florman, Samuel C. *The Civilized Engineer.* New York: St. Martin's Press, 1987.

Florman, Samuel C. *The Existential Pleasures of Engineering.* New York: St. Martin's Press, 1976.

Frankl, Viktor E. *Man's Search for Meaning.* New York: Washington Square Press, 1985.

Gleick, James. *Chaos: Making a New Science.* New York: Penguin, 1988.

Grandy, W.T. Jr. "Resource Letter ITP-1: Information Theory in Physics." *American Journal of Physics* 65 (June, 1997): 466-476.

Gribbin, John. *Q is for Quantum: An Encyclopedia of Quantum*

Mechanics. New York: Touchstone Books, 2000.

Gribbin, John. *In Search of Schrödinger's Cat: Quantum Physics and Reality.* New York: Bantam, 1984.

Griffin, David Ray (ed.). *The Reenchantment of Science.* New York: State University of New York Press, 1988.

Harrison, Edward. *Masks of the Universe* (2nd edition). Cambridge: Cambridge University Press, 2003.

Harrison, Edward. *Cosmology: The Science of the Universe.* Cambridge: Cambridge University Press, 2000.

Hobson, Art. *Physics: Concepts and Connections* (4th edition). Englewood Cliffs, New Jersey: Pearson Prentice Hall, 2006.

Hoffmann, Banesh. *Relativity and Its Roots.* New York: Scientific American Books, 1983.

Kauffman, Stuart. *Reinventing the Sacred: A New View of Science, Reason, and Religion.* New York: Basic Books, 2008.

Kauffman, Stuart. *At Home in the Universe.* Oxford: Oxford University Press, 1995.

Kitcher, Philip. *Abusing Science: The Case Against Creationism.* Cambridge, Massachusetts: MIT Press, 1987.

Jones, Roger S. *Physics for the Rest of Us: Ten Basic Ideas of Twentieth-Century Physics That Everyone Should Know...and How They Shaped Our Culture and Consciousness.* Chicago: Contemporary Books, 1992.

Leggett, A. J. *The Problems of Physics.* Oxford: Oxford University Press, 1987.

Lightman, Alan. *Great Ideas in Physics.* New York: McGraw-Hill, 1992.

Macaulay, David and Neil Ardley. *The New Way Things Work.* Boston: Houghton Mifflin, 1998.

Margulis, Lynn, Dorion Sagan and Lewis Thomas. *Microcosmos: Four Billion Years of Evolution from Our Microbial Ancestors.* New York: Summit Books, 1986.

Petroski, Henry. *To Engineer is Human.* New York: Vintage Books, 1992.

Prigogine, Ilya and Isabelle Stengers. *Order out of Chaos: Man's New Dialog with Nature.* Toronto: Bantam Books, 1984.

Raymo, Chet. *Skeptics and True Believers.* New York: Walker & Co., 1998.

Rees, D.A. "Commentary: Time for Scientists to Pay Their Dues." *Nature* 363 (20 May, 1993): 203-204.

Ridley, Matt. *Genome.* New York: HarperCollins, 2000.

Sagan, Carl. *Cosmos.* New York: Random House, 1980.

Sagan, Carl. *The Demon-Haunted World: Science as a Candle in the Dark.* New York: Random House, 1996.

Schwartz, Joseph and Michael McGuinness. *Einstein for Beginners.* New York: Pantheon Books, 1990.

Schwartz, Joseph. *The Creative Moment: How Science Made Itself Alien to Modern Culture.* New York: HarperCollins, 1992.

Senge, Peter M. *The Fifth Discipline: The Art and Practice of the*

Learning Organization. New York: Doubleday, 1994.

Shu, Frank. *The Physical Universe: An Introduction to Astronomy.* Mill Valley, California: University Science Books, 1982.

Stapp, Henry. *Mindful Universe: Quantum Mechanics and the Participating Observer.* New York: Springer, 2007.

Swimme, Brian. *The Hidden Heart of the Cosmos: Humanity and the New Story.* Maryknoll, New York: Orbis, 1996.

von Baeyer, Hans Christian. *Information: The New Language of Science.* Cambridge: Harvard University Press, 2003.

Waldrop, M. Mitchell. *Complexity: The Emerging Science at the Edge of Order and Chaos.* New York: Simon and Schuster, 1993.

Young, Louise B. *The Unfinished Universe.* New York: Simon and Schuster, 1986.

Ordering Information

To order additional copies of this book, please visit our website at http://www.scienceintegration.org.

The Science Integration Institute is a 501(c)(3) nonprofit organization. All proceeds from the sale of this book go toward developing and expanding our programs.

Thank you for your support!

Made in United States
Troutdale, OR
11/03/2026

56852543R00080